Nonlinear Differential Equations

INTERNATIONAL SERIES IN PURE AND APPLIED
MATHEMATICS

WILLIAM TED MARTIN, *Consulting Editor*

RAIMOND A. STRUBLE

PROFESSOR OF APPLIED MATHEMATICS
NORTH CAROLINA STATE COLLEGE

NONLINEAR
DIFFERENTIAL EQUATIONS

1962

McGRAW-HILL BOOK COMPANY, INC.

NEW YORK   TORONTO   LONDON

NONLINEAR DIFFERENTIAL EQUATIONS

62246

PREFACE

This book has been written, first of all, to serve as a text for a one-semester advanced undergraduate or beginning graduate course in nonlinear differential equations. However, since it was prepared with the needs of the applied mathematician, engineer, and physicist specifically in mind, it should also prove useful as a reference text for scientists and engineers working in applied fields. This is not to suggest that the book is anything but a mathematics text; rather, it is to suggest that we must recognize and then compensate for the limited mathematical experience of many who today encounter and ponder nonlinear problems. There is nothing new in this approach, but it is seldom seen in mathematical works written for those with the high level of knowledge and achievement expected here. For example, it may appear inconsistent to place side by side elementary discussions of standard mathematical notation and advanced analytical techniques, but the alternatives are either to accept a mathematically unsatisfactory and incomplete job or to continue to deny a vast audience the genuine fruits of this vital and dynamic subject.

It has been the intention of the author to provide for rapid (though modest) contact with a majority of the mathematically significant concepts of nonlinear differential equations theory without overburdening the reader with a lot of loose ends. A concerted attempt has been made for brevity of treatment (consistent with mathematically sound principles) and simplification of concepts. Thus, it has been the further intention of the author to err (if he must) by recording all too little, rather than too

v

much, and by oversimplifying, rather than overgeneralizing. The resultant shortcomings of this approach may, perhaps, be compensated by the multitude of exercises which form an integral part of the text.   These contribute limited amounts of auxiliary, though sometimes essential, material, prepare the reader for subsequent work, and provide a running criticism of the text material itself.   For the student, the last function of the exercises is by far the most important, and it is questionable if one can appreciate the real flavor of the work without careful note of this fact.   Examples appear throughout the text and in the lists of exercises.   These also contribute additional material, but more often than not, serve to illustrate the theorems, important concepts, or merely the notation and, in doing so, provide a link with more practical aspects of the theory.   Chapter 8, which consists entirely of examples, stands in marked contrast to the theoretical pattern established in the earlier chapters.   The asymptotic method illustrated therein is eminently practical and should dispel the notion that a variety of specialized techniques is required for treating traditional problems relating to linear and nonlinear oscillations.

The chapter and section titles are a sufficient indication of the total content.   Though the chapter material represents a connected account of many areas of interest, the chapters themselves are not significantly interdependent and represent more or less distinct blocks of the total structure.   A well-informed teacher may readily expand the content of any one chapter, drop or replace a chapter which, for example, might represent old material for a select audience, or insert a particular text chapter into another course or seminar.   Only a few of the more pertinent references are given throughout the text.   The list of general references includes excellent bibliographical sources, as well as other related information.

The starting point of this book was a set of lecture notes prepared for (and during) an internal seminar held at The Martin

Company, Denver Division. I wish to record here my appreciation to The Martin Company for providing me with the opportunity to participate in this seminar and with excellent secretarial help during the preparation of the lecture notes. I should like to thank John E. Fletcher and Steve M. Yionoulis for valuable assistance in the preparation of the manuscript and Mary Sue Davis for a superb job in typing the manuscript. As a student, I was fortunate to inherit from my teachers, especially Ky Fan, Joseph P. LaSalle, Karl Menger, and Arnold E. Ross, some of the rich traditions and finer things in mathematics. I sincerely hope that through this book I will share with future students at least a small part of this inheritance.

Raimond A. Struble

CONTENTS

Chapter 1

PRELIMINARY CONSIDERATIONS

1. *Linear Second-order Equations*

Consider the differential equation

$$\frac{d^2x}{dt^2} + x = 0 \tag{1}$$

which leads to the simple harmonic motion

$$x = A \sin (t + \Phi) \tag{2}$$

for arbitrary (constant) A and Φ. Let us introduce a second variable

$$y = \frac{dx}{dt} = A \cos (t + \Phi) \tag{3}$$

so that (2) and (3) together define the circle $x^2 + y^2 = A^2$ in parametric form with t as parameter. The solution in the xy plane is viewed, therefore, as a circle of radius $|A|$ centered at the origin.

A solution curve, viewed in the xy plane, is called a *trajectory*, and the xy plane itself is called the *phase plane*. A trajectory is oriented by the parameter t, and the direction of increasing t is indicated as in Fig. 1 by arrowheads. Note that from the definition of y, the arrowheads necessarily point toward positive x above the x axis and toward negative x below the x axis. A clockwise

1

motion is thus indicated.   The trivial solution of (1) corresponds to the origin $x = 0$, $y = 0$ and is called a *singular solution* or *point solution*.   It represents a position of equilibrium.   In this case there is but one position of equilibrium, and it is called a *center*, since all near trajectories are closed paths.   Closed paths generally (but not always) correspond to periodic solutions, while periodic solutions always lead to trajectories which are closed paths.

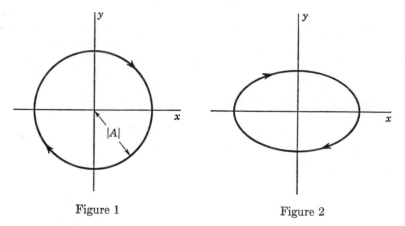

Figure 1   Figure 2

Let us now consider the trajectories defined by the equation

$$\frac{d^2x}{dt^2} + k\frac{dx}{dt} + \omega^2 x = 0 \tag{4}$$

where each of k and ω is a constant.   Without damping, i.e., $k = 0$, each trajectory of (4) is an ellipse (see Fig. 2).   However, with damping, the trajectories are modified considerably.   The nature of a solution depends upon the characteristic roots,

$$\lambda_1 = -\frac{k}{2} - \sqrt{\left(\frac{k}{2}\right)^2 - \omega^2} \qquad \lambda_2 = -\frac{k}{2} + \sqrt{\left(\frac{k}{2}\right)^2 - \omega^2}$$

We shall examine the various cases in turn.

Case 1: $\omega^2 > (k/2)^2$

Let $\omega_1 = \sqrt{\omega^2 - (k/2)^2}$ so that $\lambda_{1,2} = -k/2 \pm i\omega_1$. The general solution is well known, namely, $x = Ae^{-kt/2} \sin(\omega_1 t + \Phi)$ for arbitrary A and Φ. In this case

$$y = -\frac{k}{2} Ae^{-kt/2} \sin(\omega_1 t + \Phi) + \omega_1 Ae^{-kt/2} \cos(\omega_1 t + \Phi)$$

Let us introduce new dependent variables

$$u = \omega_1 x = \omega_1 Ae^{-kt/2} \sin(\omega_1 t + \Phi) \qquad (5)$$
$$v = y + \frac{k}{2} x = \omega_1 Ae^{-kt/2} \cos(\omega_1 t + \Phi)$$

If we interpret the solution as a trajectory in the uv plane, we obtain a spiral. Indeed from (5) we have

$$\rho^2 = u^2 + v^2 = \omega_1{}^2 A^2 e^{-kt}$$

and
$$\frac{u}{v} = \tan(\omega_1 t + \Phi)$$

Thus, for example, if $k > 0$, ρ^2 decreases monotonically as t increases, while the ratio u/v varies periodically with t (see Fig. 3). Again the motion is clockwise, although in this case v

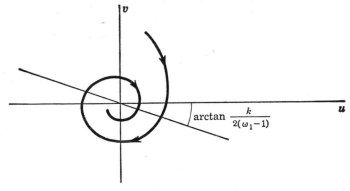

Figure 3

is not du/dt. In fact, u and v satisfy the equations

$$\frac{du}{dt} = -\frac{k}{2}u + \omega_1 v$$
$$\frac{dv}{dt} = -\omega_1 u - \frac{k}{2}v$$

(6)

For $k < 0$, the trajectories spiral clockwise away from the origin. We note that the uv origin corresponds to the xy origin, i.e., the position of equilibrium. It is called a *focus* since near trajectories spiral either to or away from it.

This rather simple picture of the trajectories has been obtained through the use of the transformation (5). The latter is a *linear transformation* of the form

$$u = b_{11}x + b_{12}y$$
$$v = b_{21}x + b_{22}y$$

(7)

If the determinant

$$\begin{vmatrix} b_{11} & b_{12} \\ b_{21} & b_{22} \end{vmatrix}$$

is different from zero, then the mapping (7) is a one-to-one mapping of the xy plane onto the uv plane. Such transformations have the following important properties:

a. The origin maps to the origin.
b. Straight lines map to straight lines.
c. Parallel lines map to parallel lines.
d. The spacings of parallel lines remain in proportion.

These hold either for the mapping from the xy plane to the uv plane or for the inverse mapping from the uv plane to the xy plane. Thus an equilateral rectangular grid work will, in general, map onto a skewed grid work with different but uniform spacing in each of the two skewed directions. Many qualitative features of the trajectories are invariant under such transformations. For example, the logarithmic spiral in Fig. 3 is the image, under the linear transformation (5), of the distorted spiral in Fig. 4.

Some quantitative information may be obtained as follows. Let us consider the linear transformation

$$u = \omega_1 x$$
$$v = \frac{k}{2} x + y \tag{8}$$

as a mapping from the two-dimensional *xy vector space* onto the two-dimensional *uv vector space*. Using rectangular cartesian

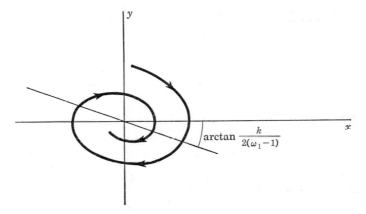

Figure 4

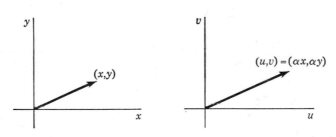

Figure 5

representation, as in Fig. 5, each *xy* vector may be identified with its end point (x,y) and its image vector under the transformation (8) by the end point (u,v) in the *uv* plane. We ask the following question: Are there any vectors in the *xy* plane which do *not* rotate under the transformation (8)? Such vectors are called *eigenvectors of the linear transformation.* If (u,v) is parallel to (x,y), the ratios v/u and y/x are equal, or what is the same, there exists a number α, called an *eigenvalue* of (8), for which

$$u = \alpha x$$
$$v = \alpha y \tag{9}$$

The eigenvalue itself is a "stretching" factor, since the length of the uv vector is $|\alpha|$ times the length of the corresponding xy vector.   But from (8) and (9) we conclude that necessarily

$$\omega_1 x = \alpha x$$
$$\frac{k}{2} x + y = \alpha y$$

or, what is the same,

$$(\omega_1 - \alpha)x = 0$$
$$\frac{k}{2} x + (1 - \alpha)y = 0 \tag{10}$$

In general, there are two nontrivial solutions of (10):

$$\alpha = 1 \qquad x = 0 \qquad (y, \text{ arbitrary}) \tag{11}$$
$$\alpha = \omega_1 \qquad \frac{k}{2} x + (1 - \omega_1)y = 0 \tag{12}$$

corresponding to the two eigenvalues $\alpha = 1$, $\alpha = \omega_1$.   If $\omega_1 = 1$, (11) and (12) are one and the same.   More generally, the first asserts that vectors parallel to the y axis are not rotated, while the second asserts that vectors with slope equal to $k/2(\omega_1 - 1)$ are not rotated.   Further, since $\alpha = 1$ in (11) and $\alpha = \omega_1$ in (12), we conclude that the lengths of the vectors parallel to the y axis remain unchanged while the lengths of the vectors with slope equal to $k/2(\omega_1 - 1)$ are stretched by the factor ω_1.   Thus the distorted spiral in Fig. 4 is obtained from the logarithmic spiral in Fig. 3 by moving the intercepts with the line $u = [k/2(\omega_1 - 1)]v$ (shown for $\omega_1 < 1$) outward in proportion to $1/\omega_1$, while leaving the intercepts on the v axis as they are.

Case 2: $(k/2)^2 > \omega^2$

Let us write the single equation (4) as the system

$$\frac{dx}{dt} = y$$
$$\frac{dy}{dt} = -\omega^2 x - ky \qquad (13)$$

In this case, we seek a linear transformation (7) such that the system (13) becomes

$$\frac{du}{dt} = \alpha_1 u$$
$$\frac{dv}{dt} = \alpha_2 v \qquad (14)$$

for suitable constants α_1, α_2. The system (14) is "uncoupled" and the solutions may be obtained immediately.

Applying the transformation (7) to (14) and using (13), we obtain

$$b_{11}y + b_{12}(-\omega^2 x - ky) = \alpha_1(b_{11}x + b_{12}y)$$
$$b_{21}y + b_{22}(-\omega^2 x - ky) = \alpha_2(b_{21}x + b_{22}y) \qquad (15)$$

If these equations are to hold identically in x and y, then the total coefficient of each of x and y must vanish. We consider, therefore, the two sets of equations

$$-\omega^2 b_{12} = \alpha_1 b_{11}$$
$$b_{11} - k b_{12} = \alpha_1 b_{12} \qquad (16)$$

and

$$-\omega^2 b_{22} = \alpha_2 b_{21}$$
$$b_{21} - k b_{22} = \alpha_2 b_{22} \qquad (17)$$

The first of (16) may be written

$$\frac{b_{12}}{b_{11}} = -\frac{\alpha_1}{\omega^2} \qquad (18)$$

and the second then becomes

$$1 + k\frac{\alpha_1}{\omega^2} = -\frac{\alpha_1^2}{\omega^2} \qquad (19)$$

Similarly, the two equations of (17) yield

$$\frac{b_{22}}{b_{21}} = -\frac{\alpha_2}{\omega^2} \tag{20}$$

and
$$1 + k\frac{\alpha_2}{\omega^2} = -\frac{\alpha_2{}^2}{\omega^2} \tag{21}$$

Equations (19) and (21) are merely versions of the characteristic equation $\lambda^2 + k\lambda + \omega^2 = 0$. Thus each of α_1 and α_2 must be a characteristic root.

With α_1 and α_2 determined, Eqs. (18) and (20) determine the ratios b_{12}/b_{11} and b_{22}/b_{21}. For convenience, we may choose $b_{11} = b_{21} = \omega^2$, and the desired linear transformation may be expressed

$$\begin{aligned} u &= \omega^2 x - \lambda_1 y \\ v &= \omega^2 x - \lambda_2 y \end{aligned} \tag{22}$$

We note that the determinant

$$\begin{vmatrix} \omega^2 & -\lambda_1 \\ \omega^2 & -\lambda_2 \end{vmatrix} = \omega^2(\lambda_1 - \lambda_2)$$

is different from zero, since $\lambda_1 \neq \lambda_2$. Thus (22) is nonsingular.

In the new variables, the solutions are given by (14) with $\alpha_1 = \lambda_1$ and $\alpha_2 = \lambda_2$. We have

$$\begin{aligned} u &= u_0 e^{\lambda_1 t} \\ v &= v_0 e^{\lambda_2 t} \end{aligned} \tag{23}$$

for arbitrary u_0 and v_0. For $k > 0$, both characteristic roots are negative and so each trajectory in the uv plane approaches the origin as $t \to \infty$. Further, the ratio

$$\frac{u}{v} = \frac{u_0}{v_0}\, e^{(\lambda_1 - \lambda_2)t}$$

approaches zero as $t \to \infty$, since $\lambda_1 - \lambda_2 = -2\sqrt{(k/2)^2 - \omega^2}$. Thus the trajectories are asymptotic to the v axis. From (23) we have

$$\frac{u^{\lambda_2}}{v^{\lambda_1}} = \frac{(u_0 e^{\lambda_1 t})^{\lambda_2}}{(v_0 e^{\lambda_2 t})^{\lambda_1}} = \frac{u_0^{\lambda_2}}{v_0^{\lambda_1}} = \text{const}$$

so that the uv trajectories lie along the curves $u = (\text{const})v^{\lambda_1/\lambda_2}$. The singular solution in this case is called a *node*. (All near trajectories tend to or away from a node without spiraling.) For $k < 0$, the solution curves are somewhat similar to those illustrated in Fig. 6. However, since both characteristic roots are

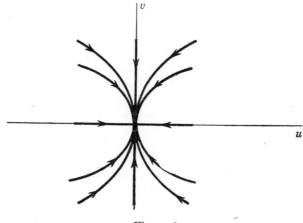

Figure 6

then positive, the arrows must be reversed and the labels on the two axes must be interchanged.

The trajectories in the phase plane are also qualitatively similar to those shown in Fig. 6 but will appear to be rotated and stretched. The eigenvalues of the transformation (22) are roots of the equation

$$\begin{vmatrix} \omega^2 - \lambda & -\lambda_1 \\ \omega^2 & -\lambda_2 - \lambda \end{vmatrix} = \lambda^2 + (\lambda_2 - \omega^2)\lambda + (\lambda_1 - \lambda_2)\omega^2 = 0$$

and the eigenvectors (i.e., invariant directions) could be obtained as before. However, in this case it is probably more important to know what happens to the u axis and v axis under (22). The

u axis maps to the line $y = \omega^2 x/\lambda_2$, while the v axis maps to the line $y = \omega^2 x/\lambda_1$. Thus the trajectories in the phase plane will appear as in Fig. 7.

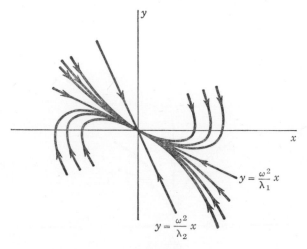

Figure 7

CASE 3: $(k/2)^2 = \omega^2$

The special case where the characteristic roots λ_1 and λ_2 are equal demands special treatment. In this case, the equations cannot be completely uncoupled. However, it is possible to find a linear transformation (7) such that one of the transformed equations becomes independent of the other. It may be solved first, and the known solution used in the remaining equation, which then becomes solvable. We use the first equation of (22) and a second equation independent of the first. Since the latter is arbitrary, except that it must be independent of the first equation, we make it simple. Let

$$u = \omega^2 x - \lambda_1 y = \left(\frac{k}{2}\right)^2 x + \frac{k}{2} y$$
$$v = \frac{k}{2} x \tag{24}$$

so that

$$\frac{du}{dt} = \left(\frac{k}{2}\right)^2 \frac{dx}{dt} + \frac{k}{2}\frac{dy}{dt} = -\frac{k}{2}\left[\left(\frac{k}{2}\right)^2 x + \frac{k}{2}y\right] = -\frac{k}{2}u$$
$$\frac{dv}{dt} = \frac{k}{2}\frac{dx}{dt} = \frac{k}{2}y = u - \left(\frac{k}{2}\right)^2 x = u - \frac{k}{2}v \tag{25}$$

Thus

$$u = u_0 e^{-kt/2} \tag{26}$$

for arbitrary u_0, and

$$\frac{dv}{dt} = u_0 e^{-kt/2} - \frac{k}{2}v \tag{27}$$

Integration of (27) yields

$$v = u_0 t e^{-kt/2} + v_0 e^{-kt/2} \tag{28}$$

for arbitrary v_0. For $k > 0$, Eqs. (26) and (28) assert that each trajectory in the uv plane approaches the origin as $t \to \infty$.

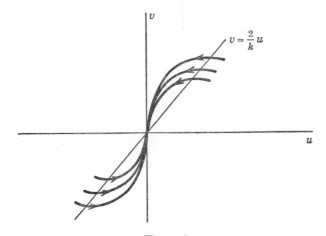

Figure 8

Further, the ratio $v/u = t + v_0/u_0$ indicates that each trajectory is asymptotic to the v axis. Using (25), we have

$$\frac{dv}{du} = \frac{dv/dt}{du/dt} = -\frac{2}{k} + \frac{v}{u}$$

Thus $dv/du = 0$ along the line $v = 2u/k$. Above this line, the tangents to the trajectories have positive slopes for $u > 0$ and negative slopes for $u < 0$, while below this line, the tangents have negative slopes for $u > 0$ and positive slopes for $u < 0$. The trajectories appear as in Fig. 8. According to (24) the line $v = 2u/k$ maps to the line $kx/2 = kx/2 + y$ or, what is the same,

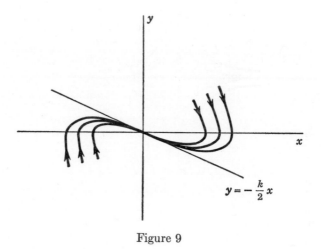

Figure 9

to the line $y = 0$. The asymptote $u = 0$ maps to the line $y = -kx/2$. Thus the trajectories in the phase plane appear as in Fig. 9. A comparison of Figs. 8 and 9 suggests that the linear transformation (24) involves a reflection as well as rotation and stretching.

CASE 4:

In order to treat the linear second-order system completely, we must consider the case when ω^2 is replaced by a negative quantity, say, $-\sigma^2$. Here we may use the results of Case 2 directly. A transformation analogous to (22) leads to a solution of the form (23). However, regardless of the sign of k, one of λ_1 and λ_2 is positive, while the other is negative. For the sake of illustration we may assume λ_2 is positive.   Then $u \to 0$ and $v \to \infty$ as $t \to \infty$, while the ratio $u/v = (u_0/v_0)e^{(\lambda_1-\lambda_2)t} \to 0$.   The trajectories in the uv plane thus appear as in Fig. 10.   The origin is called a *saddle*

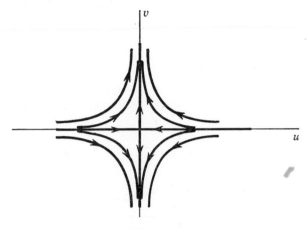

Figure 10

point for rather obvious reasons.   The u axis and v axis map to the lines

$$y = -\frac{\sigma^2}{\lambda_2} x$$

and

$$u = -\frac{\sigma^2}{\lambda_1} x$$

(29)

respectively.   Hence the phase-plane trajectories appear as in Fig. 11.

The general two-dimensional first-order system

$$\frac{dx}{dt} = ax + by$$
$$\frac{dy}{dt} = cx + ey$$

(30)

with each of a, b, c, and e constant is equivalent to the general second-order equation (4).   For example, if we introduce the variable $z = ax + by$, then (30) becomes

$$\frac{dx}{dt} = z$$

$$\frac{dz}{dt} = a\frac{dx}{dt} + b\frac{dy}{dt} = az + bcx + bey = az + bcx + e(z - ax)$$

$$= (bc - ea)x + (a + e)z$$

and so x satisfies the second-order equation

$$\frac{d^2x}{dt^2} - (a + e)\frac{dx}{dt} + (ea - bc)x = 0$$

which is of the form (4). Of course, the quantity $ea - bc$ may be negative as in Case 4 above. On the other hand, every

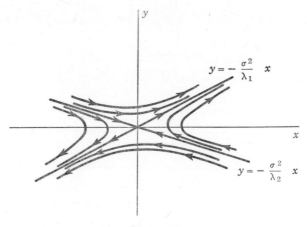

Figure 11

second-order equation of the form (4) leads to a first-order system of the form (30) with $y = dx/dt$. The parameter t is eliminated from the system (30) upon dividing the second equation by the first. The resulting equation

$$\frac{dy}{dx} = \frac{cx + ey}{ax + by}$$

then defines the direction field (or vector field) in the phase plane, and the trajectories are merely the integrals of this direction field, parameterized in the simplest and most natural manner. We note that the direction field may also be expressed in the differential form

$$\frac{dy}{cx + ey} = \frac{dx}{ax + by}$$

which readily generalizes to higher dimensions.

EXERCISES

1. Solve the equation

$$\frac{d^2x}{dt^2} + \omega^2 x = 0 \tag{i}$$

as follows. Multiply (i) by dx/dt and integrate to obtain

$$\left(\frac{dx}{dt}\right)^2 + \omega^2 x^2 = c \qquad \text{const} \geq 0 \tag{ii}$$

Thus the phase-plane trajectories (see Fig. 2) are the concentric ellipses $y^2 + \omega^2 x^2 = c$. Separate the variables in (ii), and complete the integration. Note that one obtains t as a function of x.

2. Derive (6) using (8) and (13).

3. Using (6), show that $\rho^2 = u^2 + v^2$ satisfies the differential equation $d\rho^2/dt = -k\rho^2$ and hence is given by $\rho^2 = (\text{const})e^{-kt}$.

4. Obtain the inverses of the linear transformations (7) and (8).

5. Verify the four properties of linear transformations a, b, c, and d stated in Case 1.

6. Show that an eigenvalue λ of the general linear transformation (7) satisfies the equation

$$\begin{vmatrix} b_{11} - \lambda & b_{12} \\ b_{21} & b_{22} - \lambda \end{vmatrix} = 0$$

7. Obtain the eigenvectors of the linear transformation (22).

8. Find the inverse of the linear transformation (22), and use this inverse to obtain the phase-plane solutions of (13) from (23).

9. Find the inverse of the linear transformation (24), and use this inverse to obtain the phase-plane solutions of (4) from (26) and (28).

10. Using (14), show that $w = u/v$ satisfies the differential equation $dw/dt = (\alpha_1 - \alpha_2)w$ and hence is given by

$$w = (\text{const})e^{(\alpha_1-\alpha_2)t}$$

11. Apply the method used for Case 2 to obtain the solutions for Case 1.

12. As in Case 3, adjoin to the *second* equation of (22) the equation $u = x$, and obtain differential equations for u and v which together form a system equivalent to (4). Solve the uv differential equations, invert the linear transformation, and obtain the phase-plane solutions.

13. Discuss parameterizations of the integral curves of the direction field $dy/dx = (cx + ey)/(ax + by)$ other than (30). For example, consider arclength as a parameter, and obtain appropriate equations.

14. Discuss the phase-plane trajectories of (4) for the degenerate case $\omega^2 = 0$. Notice that the origin is not a center, a node, or a focus.

15. Characterize the nature of the singular solution $x = 0$, $y = 0$ of the system (30) in terms of the coefficients a, b, c, and e.

2. *Some Nonlinear Second-order Equations*

Consider the nonlinear equation

$$\frac{d^2x}{dt^2} + \omega^2x + \beta x^3 = 0 \tag{31}$$

which is known as the Duffing equation. Normally, $|\beta|$ is a small constant and depicts a small deviation from linearity in the restoring term. If $\beta > 0$, (31) is referred to as the *hard spring case*, while if $\beta < 0$, (31) is referred to as the *soft spring case*. It is equivalent to the system

$$\frac{dx}{dt} = y$$
$$\frac{dy}{dt} = -\omega^2 x - \beta x^3$$

(32)

with the direction field given by

$$\frac{dy}{dx} = -\frac{\omega^2 x + \beta x^3}{y}$$

(33)

If the variables in (33) are separated, one obtains for the integral curves

$$y^2 + x^2\left(\omega^2 + \frac{\beta x^2}{2}\right) = c$$

(34)

where c is an integration constant.   For positive β, real solutions exist only for $c \geq 0$ and these are illustrated in Fig. 12.   The

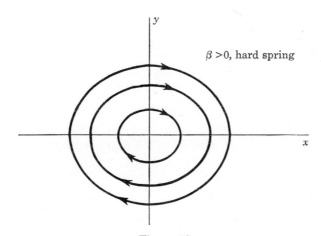

$\beta > 0$, hard spring

Figure 12

constant of integration c represents an energy level for a given trajectory.   The origin is a singular trajectory (a center), while the nontrivial solutions are represented by a family of concentric closed paths.   The latter represent periodic solutions of (31).

The trajectories for $\beta < 0$ are illustrated in Fig. 13. In this case, there are three singular solutions. These are represented by the origin and the points $y = 0$, $x = \pm(\omega^2/-\beta)^{1/2}$ on either side of the origin. The origin is a center, while each of the other two singular trajectories is a saddle point. Indeed, near the origin all trajectories are closed paths and correspond to periodic solu-

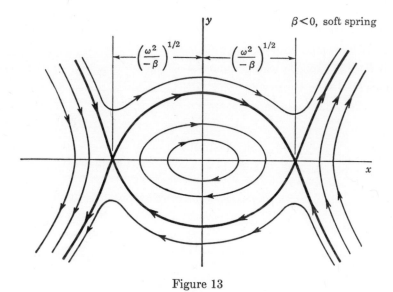

Figure 13

tions. These are very similar to the periodic solutions of Fig. 12 near the origin. The heavy loop enclosing the origin in Fig. 13 consists of two special solutions called *separatrices*. Also, there are four other separatrices which are depicted in Fig. 13 by the heavy curves on either side of the saddle points.

The final integration is reduced to a quadrature upon separation of variables in (34). Thus, we have

$$dt = \frac{dx}{\pm \sqrt{c - x^2(\omega^2 + \beta x^2/2)}} \tag{35}$$

The several cases illustrated in Figs. 12 and 13 correspond to various choices of c and β. In general, a quadrature of (35) will involve elliptic integrals and elliptic functions. Reduction to standard form in each case is left as an interesting exercise for the student.

If in (31) the parameter β is given by $\beta = -\omega^2/6$, then (31) becomes

$$\frac{d^2x}{dt^2} + \omega^2\left(x - \frac{x^3}{6}\right) = 0 \tag{36}$$

and the spring characteristic $f(x) = x - x^3/6$ is recognized as a two-term power-series expansion of $\sin x$. Thus it might be expected that those solutions with $|x|$ small would be similar to those of the equation

$$\frac{d^2x}{dt^2} + \omega^2 \sin x = 0 \tag{37}$$

Equation (37) is referred to as the *simple pendulum equation* since it depicts the idealized motions of a simple pendulum in a vacuum. The "small" oscillations of a simple pendulum are given approximately by (36). However, (37) is no more difficult to treat than its approximate (36). Equation (37) is equivalent to the system

$$\frac{dx}{dt} = y$$
$$\frac{dy}{dt} = -\omega^2 \sin x \tag{38}$$

with a direction field given by

$$\frac{dy}{dx} = -\omega^2 \frac{\sin x}{y} \tag{39}$$

If the variables in (39) are separated, one obtains for the phase-plane trajectories the equation

$$\frac{y^2}{2} - \omega^2 \cos x = c \tag{40}$$

where c is an integration constant. These trajectories are illustrated in Fig. 14.   Owing to the periodic term in (40), there are infinitely many singular trajectories.   There are centers located along the x axis at each even multiple of π and saddle points at each odd multiple of π.   The closed trajectories near each center represent the oscillatory motions of the pendulum which center about the stable equilibrium position.   The saddle points correspond to the unstable equilibrium position (the inverted pendulum), while the wavy curves in Fig. 14, appreciably

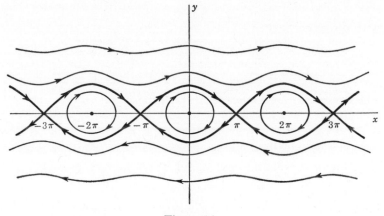

Figure 14

away from the x axis, correspond to the rotary (spinning) motions of the pendulum.   The transition from oscillatory and to rotary motion (the heavy loops in Fig. 14) is depicted by separatrices.

The final integration again reduces to a quadrature.   For upon separation of variables in (40), one obtains

$$\sqrt{2}\,dt = \frac{dx}{\pm\,\sqrt{c + \omega^2 \cos x}} \tag{41}$$

which again leads to elliptic integrals and elliptic functions.

Typical effects of damping in a nonlinear system are illustrated in the equation

$$\frac{d^2x}{dt^2} + k\,\frac{dx}{dt}\left|\frac{dx}{dt}\right| + \omega^2 \sin x = 0 \tag{42}$$

If $k > 0$, the damping acts in opposition to the motion, while the magnitude of the damping is proportional to the square of the speed. This is referred to as *Newtonian damping*. Equation (42) depicts the Newtonian damped motions of a simple pendulum. It is equivalent to the system

$$\frac{dx}{dt} = y$$
$$\frac{dy}{dt} = -\omega^2 \sin x - ky|y| \tag{43}$$

with direction field

$$\frac{dy}{dx} = -\frac{\omega^2 \sin x + ky|y|}{y} \tag{44}$$

The latter may be expressed in the form

$$\frac{dy^2}{dx} \pm 2ky^2 = -2\omega^2 \sin x \tag{45}$$

where the plus sign prevails whenever $y > 0$, and the minus sign whenever $y < 0$. Except for these changes in sign, (45) is a linear equation in the quantity y^2 with x as independent variable. Hence, the solutions of (45) are elementary. In fact, one writes immediately

$$y^2 = ce^{-2(\pm k)x} + \frac{2\omega^2}{1 + 4k^2} \cos x - \frac{4\omega^2(\pm k)}{1 + 4k^2} \sin x \tag{46}$$

where c is an integration constant, and where the ($\pm$) sign is as in (45). It is only necessary to ascertain the proper choices of the integration constant so as to "fit together" the pieces of a solution given in (46) as the sign changes become effective. This occurs in the phase plane, of course, at each point where a trajectory crosses the x axis. The trajectories defined by (46) are illustrated in Fig. 15. As in the case of an undamped pendulum, the points $y = 0$, $x = 0$, $\pm\pi$, $\pm 2\pi$, . . . represent singular solutions. The odd multiples of π correspond to saddle points, and the even multiples of π correspond to (stable) foci. Figure 15 shows that every nontrivial trajectory (other than the separatrices going into the saddle points) ultimately winds about one of the foci. Thus, the motion of the pendulum ultimately reduces to damped oscillations centered about the stable equilibrium position. The number (if any) of full rotations exhibited by the pendulum, prior to the oscillatory phase, depends upon the initial

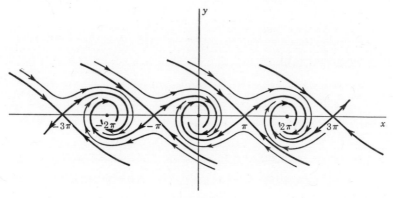

Figure 15

magnitude of $y = dx/dt$. As to be expected, the greater the initial speed, the greater the number of full rotations.

As in the previous examples, the final integration reduces to a quadrature. For upon separation of variables in (46), one obtains

$$dt = \frac{dx}{\pm \sqrt{ce^{-2(\pm k)x} + 2\omega^2/(1 + 4k^2) \cos x - [4\omega^2(\pm k)/(1 + 4k^2)] \sin x}}$$

However, in this case, a numerical quadrature is indicated.

EXERCISES

1. For $\beta > 0$ and $c > 0$, determine the two real roots of the equation

$$x^2 \left(\omega^2 + \beta \frac{x^2}{2} \right) = c \qquad \text{(i)}$$

These are the x intercepts of a phase-plane trajectory in Fig. 12 and are at a distance r from the origin. With the change of variable $x/r = \cos \theta$, show that the right member of (35) is proportional to $d\theta/\sqrt{1 - k^2 \sin^2 \theta}$, where $k^2 < 1$. Express the solution of (31) in terms of standard elliptic functions, and determine the least period of the motion.

2. For $\beta < 0$, show that (i) has real roots only for $c < -\omega^2/2\beta$. In particular, show that (i) has four real roots for $0 < c < -\omega^2/2\beta$ and two real roots for $c < 0$. Discuss the roots of (i) for $c = -\omega^2/2\beta$ and $c = 0$.

3. The trajectories appreciably away from the x axis in Fig. 13 correspond to values of the integration constant c greater than $-\omega^2/2\beta$. Explain why such trajectories cannot intersect the x axis. The closed paths in Fig. 13 and certain (explain which) of the parabolic-type trajectories on either side of the saddle points correspond to $0 < c < -\omega^2/2\beta$. Which of the trajectories of Fig. 13 correspond to $c < 0$? What values of c correspond to the separatrices of Fig. 13?

4. Express (35) in terms of standard elliptic functions for each of the several ranges of the integration constant c. Also derive closed form expressions for the separatrices.

5. Discuss the relationship between the phase-plane trajectories of Fig. 14 and the integration constant c in (40).

6. Express (41) in terms of standard elliptic functions for each of several ranges of the integration constant c. Determine the least "oscillatory" periods and the least "rotary" periods of the pendulum motion.

7. Derive (46) from (45).

8. Explain in detail how a solution of (42) is constructed using (46). In particular, assume $y = y_0$ for $x = 0$, and show how to determine the number of full rotations of the pendulum.

9. The equation

$$\frac{d^2u}{dt^2} + u = ku^2 \tag{ii}$$

where k is a small positive quantity, appears in the theory of equatorial satellite orbits of an oblate spheroid. (u depicts the variations from a constant in $1/r$, where r is the distance from the center of the spheroid to the satellite, and t is an angular variable.) Discuss the phase-plane trajectories of (ii), and express the solutions in terms of standard elliptic functions.

10. Discuss the phase-plane trajectories of the equation

$$\frac{d^2x}{dt^2} + x = \frac{a}{b - x}$$

where each of a and b is a positive constant. Consider various possibilities in regard to the relative magnitudes of a and b. (This equation depicts, approximately, the motion of a magnet suspended by a spring above a large fixed iron plate.)

11. Show that the phase-plane trajectories of the equation

$$\frac{d^2x}{dt^2} + f(x) = 0 \tag{iii}$$

are given by

$$y^2 = c - F(x) \tag{iv}$$

where $F(x) = 2 \int_0^x f(u) \, du$ and c is an integration constant. Explain in detail how the phase-plane trajectories may be constructed from a plot of $F(x)$ versus x. In terms of the geometric properties of the F curve, discuss singular points, periodic solutions, separatrices, symmetry of trajectories, extent of trajectories, etc. Illustrate by reference to the special cases of (iii) treated previously throughout this section. Discuss the integration of (iv).

12. Show that the phase-plane trajectories of the damped system

$$\frac{d^2x}{dt^2} + k\frac{dx}{dt} + f(x) = 0 \tag{v}$$

satisfy (iv) with $c = c_0 - 2k \int_0^t y^2 \, dt$, where c_0 is an integration constant. Thus c is either nondecreasing or nonincreasing along a trajectory. Explain how the trajectories of (v) for $k \neq 0$ "cut across" those of (iii). In particular, discuss the nature of the singular trajectories of (v) and show that periodic motion cannot occur if $k \neq 0$.

13. Discuss the phase-plane trajectories of a simple pendulum with linear (viscous) damping. Refer to Fig. 14 and Exercise 12.

14. Extend the results in Exercise 12 to cover the general nonlinear second-order equation

$$\frac{d^2x}{dt^2} + g\left(x,\frac{dx}{dt},t\right) + f(x) = 0$$

for which either $yg(x,y,t) \geq 0$ or $yg(x,y,t) \leq 0$ throughout. Illustrate by verifying the qualitative features of the trajectories of (42) as depicted in Fig. 15.

3. *The Initial-value Problem*

The phase-plane concept is introduced as the natural setting for formulating the initial-value problem. The initial-value problem for a second-order equation is as follows: Find a solution $x(t)$ which for $t = t_0$ satisfies

$$x = x_0$$
$$\frac{dx}{dt} = y_0 \tag{47}$$

In the phase plane, (47) specifies a point $P_0 = (x_0,y_0)$, and the initial-value problem is that of finding a trajectory which passes through the point P_0 for $t = t_0$. For each system discussed in the previous two sections, there would be no loss in generality to assume that $t_0 = 0$ (i.e., we may choose to parameterize the trajectory so that $t = 0$ corresponds to P_0). Clearly P_0, in general, is neither the beginning nor the end of a trajectory but merely a point on a trajectory. An important exception to this is the singular or point trajectory. For the systems discussed previously and for any point P_0, the solution of the initial-value problem exists and is unique, whether singular or not. This is to say, if $P_0 = (x_0,y_0)$ is a point in the phase plane, there is one and only one trajectory Γ, such that P_0 is a point of Γ. For example, if P_0 is a singular point of the system, then Γ consists of the single point P_0. One should note the important distinction made here between a trajectory as a parameterized solution of the differential system and as a plane curve which may appear to "pass through" a singular point.

There are many excellent reasons for introducing the phase-plane, or what we shall presently generalize to phase-space, concept. In the first place, every differential system treated is expressed as a first-order vector system and the treatment thus becomes unified. Pure geometric and vector concepts may be brought to bear at any time as an aid to the understanding of the algebraic and analytical processes. Further, all variables are treated on a par so that there is no artificial distinction made between the initial values of the various dependent variables. For example, the initial value of dx/dt in (47) plays the same role in the initial-value problem as does the initial value of x. Indeed, the solution is here thought of as a *pair* of functions $x(t)$ and $y(t)$ for which $x(t_0) = x_0$ and $y(t_0) = y_0$, rather than as a single function $x(t)$ for which $x = x_0$ and $dx/dt = y_0$ when $t = t_0$.

One should note that the initial-value problem is fundamentally a "local" problem in the sense that it is concerned with a trajectory Γ near a single point, the initial point. To be sure, there are important questions to be answered in regard to the over-all nature of Γ, as to where it goes or from where it comes, but the foremost questions, those of existence and uniqueness, are each of a strictly local nature.

EXERCISES

1. Consider the system

$$\frac{dx}{dt} = -x$$

$$\frac{dy}{dt} = -y$$

(i)

with direction field

$$\frac{dy}{dx} = \frac{y}{x} \tag{ii}$$

The "solutions" of (ii) are usually expressed in the form $y = cx$ for arbitrary c.   Explain why this is but a half-truth, and discuss the initial-value problem for (i).

2. Discuss the initial-value problem for a system whose direction field is given by a separable equation of the form

$$N(x) \, dx - M(y) \, dy = 0$$

3. Discuss the initial-value problem for a system whose direction field is given by an exact equation of the form

$$N(x,y) \, dx - M(x,y) \, dy = 0 \tag{iii}$$

4. Discuss the relationship between integrating factors of (iii) and parameterizations of the integral curves of the direction field $dy/dx = N(x,y)/M(x,y)$.

5. Consider the general second-order equation

$$\frac{d^2x}{dt^2} + f\left(x, \frac{dx}{dt}\right) = 0 \tag{iv}$$

If Γ is a phase-plane trajectory of (iv) through the point P_0, then express t as a line integral along the curve Γ in three different forms.   Express (iv) as a system of two equations and use, in turn, each of the equations and then both.   Generalize to a system with direction field given by (iii).

6. A *boundary-value problem* for (iv) might read: Find a solution of (iv) which satisfies the two linear relations

$$a_1 x + b_1 \frac{dx}{dt} = c_1 \qquad \text{for } t = t_1$$

$$a_2 x + b_2 \frac{dx}{dt} = c_2 \qquad \text{for } t = t_2 \tag{v}$$

where each of $a_1, b_1, c_1, a_2, b_2, c_2, t_1, t_2$ is given.   For example, one might seek a solution $x(t)$ which vanishes for $t = 0$ and $t = 1$.

Rephrase the boundary-value problem (v) in geometrical terms with reference to the phase plane of (iv).

4. *Higher-order Systems*

The system (30) may be given the following vectorial interpretation. Let us define a column vector

$$\bar{x} = \begin{pmatrix} x \\ y \end{pmatrix}$$

and, when x and y are differentiable functions of a scalar variable t, the derivative

$$\frac{d\bar{x}}{dt} = \begin{pmatrix} \dfrac{dx}{dt} \\ \dfrac{dy}{dt} \end{pmatrix}$$

Then (30) is equivalent to the vector equation

$$\frac{d\bar{x}}{dt} = A\bar{x}$$

where A is the 2-by-2 matrix

$$A = \begin{pmatrix} a & b \\ c & e \end{pmatrix}$$

and where $A\bar{x}$ denotes the usual row by column multiplication of a square matrix and a column vector.

More generally, if we consider an n-dimensional system

$$\frac{dx_1}{dt} = a_{11}x_1 + a_{12}x_2 + \cdots + a_{1n}x_n$$

$$\frac{dx_2}{dt} = a_{21}x_1 + a_{22}x_2 + \cdots + a_{2n}x_n$$

$$\cdots \cdots \cdots \cdots \cdots \cdots \cdots \cdots$$

$$\frac{dx_n}{dt} = a_{n1}x_1 + a_{n2}x_2 + \cdots + a_{nn}x_n$$

then with $\quad \bar{x} = \begin{pmatrix} x_1 \\ x_2 \\ \cdots \\ x_n \end{pmatrix} \quad$ and $\quad \dfrac{d\bar{x}}{dt} = \begin{pmatrix} \dfrac{dx_1}{dt} \\ \dfrac{dx_2}{dt} \\ \cdots \\ \dfrac{dx_n}{dt} \end{pmatrix}$

one obtains the vector equation

$$\frac{d\bar{x}}{dt} = A\bar{x} \tag{48}$$

where now A is the n-by-n matrix

$$\begin{pmatrix} a_{11} & a_{12} & \cdots & a_{1n} \\ a_{21} & a_{22} & \cdots & a_{2n} \\ \cdots & \cdots & \cdots & \cdots \\ a_{n1} & a_{n2} & \cdots & a_{nn} \end{pmatrix}$$

Equation (48) is the vectorial version of the most general linear homogeneous system of differential equations. The n-dimensional vector space is called the *phase space*, and a solution or trajectory is a space curve in the n-dimensional phase space.

EXAMPLE 1

Consider the third-order equation

$$4\frac{d^3x}{dt^3} + 3\frac{d^2x}{dt^2} + 2\frac{dx}{dt} + x = 0 \tag{49}$$

We define

$$\bar{x} = \begin{pmatrix} x_1 \\ x_2 \\ x_3 \end{pmatrix} = \begin{pmatrix} x \\ \dfrac{dx}{dt} \\ \dfrac{d^2x}{dt^2} \end{pmatrix}$$

and note that (49) becomes

$$\frac{dx_3}{dt} = -\frac{x_1}{4} - \frac{x_2}{2} - \frac{3x_3}{4}$$

upon solving for the highest derivative. Hence (49) is equivalent to the third-order system

$$\frac{dx_1}{dt} = x_2$$

$$\frac{dx_2}{dt} = x_3 \tag{50}$$

$$\frac{dx_3}{dt} = -\frac{x_1}{4} - \frac{x_2}{2} - \frac{3x_3}{4}$$

where the first two merely define the notation. The vectorial version of (50) is (48) with

$$A = \begin{pmatrix} 0 & 1 & 0 \\ 0 & 0 & 1 \\ -\frac{1}{4} & -\frac{1}{2} & -\frac{3}{4} \end{pmatrix}$$

In three-space, the trajectories are integral curves of the system

$$\frac{dx_1}{x_2} = \frac{dx_2}{x_3} = \frac{-dx_3}{x_1/4 + x_2/2 + 3x_3/4}$$

This system defines a three-dimensional direction field where the vectorial direction at the point (x_1, x_2, x_3) is given by the direction numbers $\left(x_2, x_3, -\frac{x_1}{4} - \frac{x_2}{2} - \frac{3x_3}{4} \right)$.

EXAMPLE 2

The nth-order equation

$$\frac{d^n x}{dt^n} + b_n \frac{d^{n-1}x}{dt^{n-1}} + b_{n-1}\frac{d^{n-2}x}{dt^{n-2}} + \cdots + b_1 x = 0$$

is equivalent to the vector equation (48)

with
$$\bar{x} = \begin{pmatrix} x_1 \\ x_2 \\ x_3 \\ \cdots \\ x_n \end{pmatrix} = \begin{vmatrix} x \\ \dfrac{dx}{dt} \\ \dfrac{d^2 x}{dt^2} \\ \cdots \\ \dfrac{d^{n-1}x}{dt^{n-1}} \end{vmatrix}$$

$$\text{and} \quad A = \begin{pmatrix} 0 & 1 & 0 & \cdots & \cdots & \cdots & \cdots & 0 \\ 0 & 0 & 1 & 0 & \cdots & \cdots & \cdots & 0 \\ 0 & 0 & 0 & 1 & 0 & \cdots & \cdots & 0 \\ \cdots & \cdots & \cdots & \cdots & \cdots & \cdots & \cdots & \cdots \\ 0 & 0 & 0 & \cdots & \cdots & \cdots & 0 & 1 \\ -b_1 & -b_2 & \cdots & \cdots & \cdots & \cdots & \cdots & -b_n \end{pmatrix}$$

Example 3

Consider the following system of two coupled second-order equations.

$$\frac{d^2x}{dt^2} + 5\frac{dy}{dt} + x - 2y = 0$$
$$\frac{d^2y}{dt^2} + 2\frac{dy}{dt} - 3x + y = 0$$
(51)

Let $x = x_1$, $dx/dt = x_2$, $y = x_3$, and $dy/dt = x_4$. Then the system (51) is equivalent to the vector equation (48) with

$$A = \begin{pmatrix} 0 & 1 & 0 & 0 \\ -1 & 0 & 2 & -5 \\ 0 & 0 & 0 & 1 \\ 3 & 0 & -1 & -2 \end{pmatrix}$$

Example 4

The linear differential equation

$$\frac{d^2x}{dt^2} + a(t)\frac{dx}{dt} + b(t)\,x = f(t)$$
(52)

is equivalent to the vector equation

$$\frac{d\bar{x}}{dt} = A(t)\,\bar{x} + \bar{f}(t)$$
(53)

where
$$A(t) = \begin{pmatrix} 0 & 1 \\ -b(t) & -a(t) \end{pmatrix}$$

and
$$\bar{f}(t) = \begin{pmatrix} 0 \\ f(t) \end{pmatrix}$$

provided
$$\bar{x} = \begin{pmatrix} x_1 \\ x_2 \end{pmatrix} = \begin{pmatrix} x \\ \dfrac{dx}{dt} \end{pmatrix}$$

The vector equation (53) represents the most general linear system of dimension two.

EXAMPLE 5

Consider once again Eq. (52).   Let us define $x = x_1$, $dx/dt = x_2$, and $t = x_3$.   Then (52) becomes

$$\frac{dx_2}{dt} = -b(x_3)\, x_1 - a(x_3)\, x_2 + f(x_3)$$

The right-hand side defines a single function of three variables x_1, x_2, and x_3, say

$$f_2(x_1,\, x_2,\, x_3) \equiv -b(x_3)\, x_1 - a(x_3)\, x_2 + f(x_3)$$

For completeness we define also two other functions

$$f_1(x_1,\, x_2,\, x_3) \equiv x_2 \qquad f_3(x_1,\, x_2,\, x_3) \equiv 1$$

Then (52) is equivalent to the system

$$\frac{dx_1}{dt} = f_1(x_1,\, x_2,\, x_3)$$

$$\frac{dx_2}{dt} = f_2(x_1,\, x_2,\, x_3)$$

$$\frac{dx_3}{dt} = f_3(x_1,\, x_2,\, x_3)$$

These equations are no longer linear in the three variables x_1, x_2, and x_3, but yet we may express the system vectorially in the form

$$\frac{d\bar{x}}{dt} = \bar{f}(\bar{x})$$

where $\qquad \bar{x} = \begin{pmatrix} x_1 \\ x_2 \\ x_3 \end{pmatrix} \qquad$ and $\qquad \bar{f}(\bar{x}) = \begin{pmatrix} f_1(\bar{x}) \\ f_2(\bar{x}) \\ f_3(\bar{x}) \end{pmatrix}$

Here we use the notation $f_k(\bar{x}) = f_k(x_1,\, x_2,\, x_3)$ to denote a scalar function of three scalar quantities x_1, x_2, and x_3 or what is the same, to denote a scalar function of the vector $\bar{x}$.   On the other hand, $\bar{f}(\bar{x})$ denotes a *vector* function of the vector $\bar{x}$.

◆ ◆ ◆ ◆ ◆ ◆ ◆

The most general n-dimensional system we shall consider may be expressed in the vector form

$$\frac{d\bar{x}}{dt} = \bar{f}(\bar{x},t) \tag{54}$$

where

$$\bar{f}(\bar{x},t) = \begin{pmatrix} f_1(\bar{x},t) \\ f_2(\bar{x},t) \\ \cdot \cdot \cdot \\ f_n(\bar{x},t) \end{pmatrix} \tag{55}$$

and

$$f_k(\bar{x},t) = f_k(x_1, x_2, \ldots, x_n, t) \qquad k = 1, 2, \ldots, n \tag{56}$$

The vector function (55) will be called the *right-hand member* and generally will be a function of the independent variable t and the n dependent variables $x_1, x_2, \ldots, x_n$. The system (54) is *linear* if each of the scalar functions (56) is linear in each of $x_1, x_2, \ldots, x_n$. The initial-value problem associated with (54) concerns a solution of (54) satisfying $\bar{x}(t_0) = \bar{c}$, for a given $\bar{c}$ and t_0. A solution vector function $\bar{x}(t)$ is called a *trajectory* and defines, parametrically, a space curve in the n-dimensional space. A *singular solution* (or point solution) is a trajectory consisting of a single point $\bar{c}$ such that $\bar{x}(t) = \bar{c}$ satisfies (54) identically in t (or possibly for $t_0 \leq t \leq t_0 + b$, for some $b > 0$). Clearly, $\bar{c}$ must satisfy

$$\bar{f}(\bar{c},t) \equiv \bar{0} \tag{57}$$

i.e., the right-hand member must vanish identically for $\bar{x} = \bar{c}$. A solution $\bar{c}$ of (57) is called a *singular point* of (54). A singular point represents a state of equilibrium of the system. A point of the phase space which is not a singular point of (54) is called a *regular point*.

It may be shown that $f_k(x_1, x_2, \ldots, x_n, t)$ is continuous and linear in $x_1, x_2, \ldots, x_n$ if and only if there are $n + 1$ continuous functions $a_{k1}, a_{k2}, \ldots, a_{kn}, f_k$ of t such that

$$f_k(x_1, x_2, \ldots, x_n, t) = a_{k1}x_1 + a_{k2}x_2 + \cdots + a_{kn}x_n + f_k$$

holds identically in $x_1, x_2, \ldots, x_n$, and t. Thus (54) is continuous and linear If and only if there exists a continuous matrix function of t,

$$A(t) = \begin{pmatrix} a_{11}(t) & a_{12}(t) & \cdots & a_{1n}(t) \\ a_{21}(t) & a_{22}(t) & \cdots & a_{2n}(t) \\ \cdots & \cdots & \cdots & \cdots \\ a_{n1}(t) & a_{n2}(t) & \cdots & a_{nn}(t) \end{pmatrix}$$

and a continuous vector function of t,

$$\bar{f}(t) = \begin{pmatrix} f_1(t) \\ f_2(t) \\ \cdots \\ f_n(t) \end{pmatrix}$$

such that $\bar{f}(\bar{x},t) = A(t)\,\bar{x} + \bar{f}(t)$ holds identically in $x_1, x_2, \ldots,$ x_n, and t. Therefore, the vector equation

$$\frac{d\bar{x}}{dt} = A(t)\,\bar{x} + \bar{f}(t)$$

represents the most general linear system of order n. The matrix function $A(t)$ is called the *coefficient matrix*, and the vector function $\bar{f}(t)$ is called the *forcing function*. A linear system with constant coefficients is one for which the coefficient matrix is a constant matrix. A linear system with periodic coefficients of period τ is one for which $A(t + \tau) = A(t)$ holds identically in t; i.e., the coefficient matrix is periodic with period τ. When the forcing function is zero, the system is said to be *homogeneous*. The origin, i.e., $\bar{x} = \bar{0}$, is a singular point of each linear homogeneous system.

When the right-hand member in (54) is independent of t, the system (or vector equation) is said to be *autonomous*. By adding one more dependent variable and one trivial differential equation (see Example 5), one may always construct an autonomous system which is equivalent to (54). By this artifice we unify our treatment, though there are instances in which one must insist upon retaining the nonautonomous form.

5. *The Taxicab Geometry*[1]

The *norm* of a vector $\bar{x}$ is defined as the scalar quantity

$$\left|\bar{x}\right| = \sum_{i=1}^{n} |x_i| \tag{58}$$

In analytical work, it is simpler to deal with the norm of a vector than with the more familiar Euclidean length

$$|\bar{x}| = \sqrt{x_1^2 + x_2^2 + \cdots + x_n^2}$$

On the other hand, the norm of a vector is clearly a measure of the magnitude of a vector, since

a. $\left|\bar{x}\right| = 0$ if and only if $\bar{x} = \bar{0}$.

b. $\lim \left|\bar{x}\right| = 0$ if and only if $\lim |\bar{x}| = 0$.

In fact, this particular norm has been given the very descriptive name *taxicab length* or *distance*, since relative to this norm, distance is accrued along rectangular (i.e. piece-wise rectilinear) paths.   Distance in the taxicab geometry is that familiar urban measurement in "block" units.   Property *a* merely states that points in n-space are distinguished by the norm.   More generally, the distance (taxicab distance or, if you prefer, the number of blocks) from

$$\bar{x} = \begin{pmatrix} x_1 \\ x_2 \\ \cdots \\ x_n \end{pmatrix} \quad \text{to} \quad \bar{y} = \begin{pmatrix} y_1 \\ y_2 \\ \cdots \\ y_n \end{pmatrix}$$

is the norm of the difference vector $\bar{x} - \bar{y}$; that is,

$$\left|\bar{x} - \bar{y}\right| = \sum_{i=1}^{n} |x_i - y_i|$$

[1] The name is due to K. Menger, "You Will Like Geometry," p. 5, Guidebook for Illinois Institute of Technology Geometry Exhibit, Museum of Science and Industry, Chicago, Ill., 1952.

Thus by property a, $\bar{x}$ and $\bar{y}$ are position vectors to different points if and only if $\|\bar{x} - \bar{y}\| \neq 0$.

Property b states that the topology induced (defined) by the norm is equivalent to the familiar Euclidean topology. This may be illustrated by considering, a little more in detail, the two geometries. In Euclidean geometry, the equation $|\bar{x}| = 1$ defines the n sphere of radius one, centered at the origin. For $n = 2$, this is a circle. In the taxicab geometry, the analogous equation $\|\bar{x}\| = 1$ defines an inscribed square (see Fig. 16). The

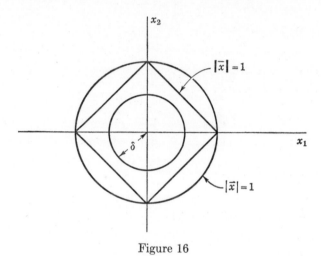

Figure 16

points inside the circle satisfy the inequality $|\bar{x}| < 1$, while the points inside the square satisfy the inequality $\|\bar{x}\| < 1$. Of course, the latter also lie inside the unit circle. This is to say, if $\|\bar{x}\| < 1$ then $|\bar{x}| < 1$. On the other hand, there exists a circle of radius δ, less than unity (any radius $\delta \leq 1/\sqrt{2}$ will do), such that the interior of the circle of radius δ, centered at the origin, lies within the square (see Fig. 16). This is to say, if $|\bar{x}| < \delta$, then $\|\bar{x}\| < 1$. What is really important for the analysis is merely the fact that inside each circle is a square, and that inside each square is a circle. Thus the circle shrinks to the origin if

and only if the square shrinks to the origin.   This is property *b*.
Other properties of the norm which we shall need are:

c. $\|\bar{x} + \bar{y}\| \leq \|\bar{x}\| + \|\bar{y}\|$ (triangular inequality).
d. $\|c\bar{x}\| = |c|\|\bar{x}\|$ where c is a scalar.
e. $\left\|\int_{t_1}^{t_2} \bar{x}(t)\, dt\right\| \leq \left|\int_{t_1}^{t_2} \|\bar{x}(t)\|\, dt\right|$

where
$$\int_{t_1}^{t_2} \bar{x}(t)\, dt = \begin{pmatrix} \int_{t_1}^{t_2} x_1(t)\, dt \\ \int_{t_1}^{t_2} x_2(t)\, dt \\ \cdots \\ \int_{t_1}^{t_2} x_n(t)\, dt \end{pmatrix}$$

We shall also have occasion to use the norm of a square matrix
A which we define similarly.

$$\|A\| = \sum_{\substack{i=1 \\ j=1}}^{n} |a_{ij}| \tag{59}$$

The following properties are easily verified.

f. $\|A\| = 0$ if and only if $\left[\sum_{\substack{i=1 \\ j-1}}^{n} (a_{ij})^2\right]^{\frac{1}{2}} = 0$.

g. $\lim \|A\| = 0$ if and only if $\lim \left[\sum_{\substack{i=1 \\ j=1}}^{n} (a_{ij})^2\right]^{\frac{1}{2}} = 0$.

h. $\|A + B\| \leq \|A\| + \|B\|$.
i. $\|cA\| \leq |c|\|A\|$ where c is a scalar.
j. $\|A\bar{x}\| \leq \|A\|\|\bar{x}\|$.
k. $\|AB\| \leq \|A\|\|B\|$.
l. $\left\|\int_{t_1}^{t_2} A(t)\, dt\right\| \leq \left|\int_{t_1}^{t_2} \|A(t)\|\, dt\right|$

where $\int_{t_1}^{t_2} A(t)\, dt = \left(\int_{t_1}^{t_2} a_{ij}(t)\, dt\right)$.
We define limiting processes for vectors and matrices as one
would expect.   For example,

$m.$ $\bar{x} = \lim\limits_{k \to \infty} \bar{x}^{(k)}$ means: For each $\epsilon > 0$, there exists a positive integer K such that $k \geq K$ implies $\left| \bar{x} - \bar{x}^{(k)} \right| < \epsilon.$

$n.$ $\bar{y} = \lim\limits_{t \to t_0} \bar{x}(t)$ means: For each $\epsilon > 0$, there exists a $\delta > 0$ such that $0 < |t - t_0| < \delta$ implies $\left| \bar{y} - \bar{x}(t) \right| < \epsilon.$

Using properties m and n, and similar limit definitions for matrices, one can show that

$o.$ $\lim\limits_{k \to \infty} \bar{x}^{(k)} = \bar{x}$ if and only if

$$\bar{x} = \begin{pmatrix} \lim\limits_{k \to \infty} x_1^{(k)} \\ \lim\limits_{k \to \infty} x_2^{(k)} \\ \cdot \ \cdot \ \cdot \\ \lim\limits_{k \to \infty} x_n^{(k)} \end{pmatrix}$$

$p.$ $\dfrac{d\bar{x}}{dt} = \lim\limits_{t_1 \to t} \dfrac{\bar{x}(t_1) - \bar{x}(t)}{t_1 - t}.$

$q.$ $\dfrac{d(AB)}{dt} = \dfrac{dA}{dt} B + A \dfrac{dB}{dt}.$

$r.$ $\dfrac{d(A\bar{x})}{dt} = \dfrac{dA}{dt} \bar{x} + A \dfrac{d\bar{x}}{dt}.$

$s.$ $\dfrac{d(A^{-1})}{dt} = - A^{-1} \dfrac{dA}{dt} A^{-1}.$

6. *Continuous, Differentiable, Analytic, and Lipschitz Vector Functions*

A vector function $\bar{f}(\bar{x},t)$ of the vector $\bar{x}$ and scalar t is *continuous in the pair* $(\bar{x},t)$ if it is continuous in the $n + 1$ variables x_1, $x_2, \ldots, x_n$, t. The word *continuous* in the above definition may be replaced, in turn, by the words *differentiable* and *analytic*. The latter might need some explanation. A function is analytic in $x_1, x_2, \ldots, x_n$, t if it can be expressed as a power series in the $n + 1$ variables $x_1, x_2, \ldots, x_n, t$.

A vector function $\bar{f}(\bar{x},t)$ is a *Lipschitz function* in $\bar{x}$ if there exists a constant m such that

$$\left| \bar{f}(\bar{x},t) - \bar{f}(\bar{y},t) \right| \leq m \left| \bar{x} - \bar{y} \right|$$

A number m for which the above is true is called a *Lipschitz constant*.

In the above definitions the ranges of the variables have not been mentioned. In most applications, each concept will concern a local characteristic of the function. It will depict a property of $\bar{f}(\bar{x},t)$ for a restricted (generally small) region of $(\bar{x},t)$ space.

Finally, we note that if $\bar{f}(\bar{x},t)$ has bounded first-partial derivatives in each of $x_1, x_2, \ldots, x_n$ (uniform in t), then $\bar{f}(\bar{x},t)$ is a Lipschitz function. This follows from the mean-value theorem for functions of several variables.

EXERCISES

1. Show that $f(x_1, x_2, \ldots, x_n, t)$ is continuous and linear in $x_1, x_2, \ldots, x_n$ if and only if there are $n + 1$ continuous functions $a_1, a_2, \ldots, a_n, f_1$ such that

$$f(x_1, x_2, \ldots, x_n, t) = a_1 x_1 + a_2 x_2 + \cdots + a_n x_n + f_1$$

holds identically in $x_1, x_2, \ldots, x_n$, and t.

2. Using the definitions (58) and (59), verify the 12 properties of the norm, a through l.

3. Using the definitions m and n, verify properties o and p.

4. Define the "double norm" of a vector $\bar{x}$ as the scalar quantity $\|\bar{x}\| = \max (|x_1|, |x_2|, \ldots, |x_n|)$. Show that in two-space, the locus $\|\bar{x}\| = 1$ is a square which circumscribes the circle $|\bar{x}| = 1$. Thus the double-norm equivalent of a Euclidean circle is also a square. The author is indebted to one of his students for suggesting the picturesque name *squircle*, a "square circle." On the other hand, the orientation of the "single-norm" square $|\bar{x}| = 1$ suggests a diamond, and so one might coin for it the name *dircle*, a "diamond circle." Illustrate, by a sketch, the

geometric relationships between circles, dircles, and squircles, and show that $\lim |\bar{x}| = 0$, if and only if $\lim |\![\,\bar{x}\,]\!| = 0$, if and only if $\lim \|\bar{x}\| = 0$. One might use the double norm in place of the norm or Euclidean length and obtain the same limit concepts. In fact, the double norm, typically, is the basis for elementary studies in analysis which are usually referred to as "advanced calculus." Explain the connection.

5. The continuous vector function $\bar{f}(\bar{x},t)$ is said to be *linear in* $\bar{x}$ if $\bar{f}(\alpha\bar{x} + \beta\bar{y}) = \alpha\bar{f}(\bar{x},t) + \beta\bar{f}(\bar{y},t)$ for all vectors $\bar{x}$, $\bar{y}$ and all scalars α, β, and t. Show that this definition of "linearity" is equivalent to that given for the right-hand member in (54).

6. Show that $\bar{f}(\bar{x},t)$ is a Lipschitz function if it is continuous in the pair $(\bar{x},t)$ and linear in $\bar{x}$.

7. Give examples of continuous functions which are not Lipschitz functions.

8. Give examples of Lipschitz functions for which "universal" (in contrast to local) Lipschitz constants exist.

9. Under what circumstance may a linear homogeneous system possess a singular point other than the origin? In what sense is such a system degenerate?

Chapter 2

THE EXISTENCE AND THE UNIQUENESS OF A SOLUTION OF THE INITIAL-VALUE PROBLEM

1. The Cauchy-Lipschitz Existence Theorem (Method of Successive Approximations)

The initial-value problem for systems of ordinary differential equations was defined in Chap. 1. Here we shall state and prove a few basic theorems concerning this problem, leaving the bulk of the interpretations and applications to subsequent chapters.

Let R denote the subset of n-space consisting of all $\bar{x}$ satisfying

$$\left| \bar{x} - \bar{c} \right| \leq a \tag{1}$$

where $\bar{c}$ is a point of n-space, $a > 0$, each fixed throughout this chapter. Let $\bar{f}(\bar{x})$ be continuous in R, and consider the following differential equation:

$$\frac{d\bar{x}}{dt} = \bar{f}(\bar{x}) \tag{2}$$

We seek a solution of (2), subject to the initial condition

$$\bar{x} = \bar{c} \qquad \text{for } t = 0 \tag{3}$$

which exists for t in an interval $0 \leq t \leq b$ for some $b > 0$, and for which x remains in R. We shall call this the *forward problem*.

In the *backward problem* one seeks a solution on an interval $-b \leq t \leq 0$ for $b > 0$. However, replacing t by $-t$ in (2) converts a backward problem to a forward problem with the right-hand member $-\bar{f}(\bar{x})$. Thus it is sufficient to direct all our attention to the forward problem. We remark that there is no loss of generality in assuming that the initial value $\bar{c}$ is given for $t = 0$, nor that (2) is an autonomous system. The forward initial-value problem is equivalent to the problem of finding a solution of the integral equation

$$\bar{x}(t) = \bar{c} + \int_0^t \bar{f}(\bar{x}(s))\, ds \tag{4}$$

for $0 \leq t \leq b$. The latter, however, is more amenable to analysis.

It is instructive to consider a solution of (4) as a fixed point of a transformation. To this end, we define, for each continuous vector function $\bar{x}(t)$ (defined on $0 \leq t \leq b$, with values in R) the transform

$$T(\bar{x}) = \bar{c} + \int_0^t \bar{f}(\bar{x}(s))\, ds \tag{5}$$

Then
$$\bar{y} = T(\bar{x}) \tag{6}$$

is a continuous (vector-valued) function of t and for sufficiently small t remains in R. Since the function $\bar{f}(\bar{x})$ is continuous, it is necessarily bounded in R. Hence let K be such that

$$\left| \bar{f}(\bar{x}) \right| \leq K \tag{7}$$

for each $\bar{x}$ in R. Then from (5) and (6), we obtain

$$\left| \bar{y} - \bar{c} \right| = \left| T(\bar{x}) - \bar{c} \right| \leq \int_0^t \left| \bar{f}(\bar{x}(s)) \right| ds \leq Kt \tag{8}$$

so long as $\bar{x}(s)$ is in R. If we let

$$b = \frac{a}{K} \tag{9}$$

then (8) implies that $\left| \bar{y} - \bar{c} \right| \leq a$ for $0 \leq t \leq b$ so long as $\bar{x}$ satisfies $\left| \bar{x} - \bar{c} \right| \leq a$ for $0 \leq t \leq b$. This is to say that the set S, consisting of all continuous functions $\bar{x}$ with values in R for

$0 \leq t \leq b$, is mapped into itself by the transformation (6). S is called an *invariant set of the transformation*.   A solution of the integral equation (4) is a fixed point of T, i.e., a solution of $\bar{x} = T(\bar{x})$, and clearly belongs to S.   If it can be shown that there exists at least one fixed point of T then a solution of the initial-value problem exists.   If T possesses exactly one fixed point, the solution of the initial-value problem is unique.

It is typical in fixed-point problems to consider iterative procedures.   One selects some member of S (arbitrarily), say $\bar{x}^{(0)}$, and defines recursively

$$\bar{x}^{(1)} = T(\bar{x}^{(0)}) \quad \bar{x}^{(2)} = T(\bar{x}^{(1)}) \quad \cdots \quad \bar{x}^{(k+1)} = T(\bar{x}^{(k)}) \quad \cdots$$

The sequence $\bar{x}^{(0)}$, $\bar{x}^{(1)}$, . . . generally wanders about in S and may or may not "converge." Numerous devices might be employed to increase the chances of convergence, improve on the starting point $\bar{x}^{(0)}$, average at each step or over several steps, etc. In the present case, the iterates are known as *successive approximations*, from which the process receives its name.   Typically, one chooses as a first approximation $\bar{x}^{(0)} \equiv \bar{c}$, i.e., the constant initial value itself, although this is not essential and often represents a very poor over-all approximation.   We have the following important theorem.

THEOREM 1

Let $\bar{x}^{(0)}$ be in S, i.e., let $\bar{x}^{(0)}$ be continuous and satisfy $\left| \bar{x}^{(0)} - \bar{c} \right| \leq a$ for $0 \leq t \leq b$.   For $k = 0, 1, \ldots$ define (by induction)

$$\bar{x}^{(k+1)} = T(\bar{x}^{(k)}) = \bar{c} + \int_0^t \bar{f}(\bar{x}^{(k)}(s))\, ds \tag{10}$$

If $\bar{f}(\bar{x})$ is a Lipschitz function, i.e., if there exists a constant m such that

$$\left| \bar{f}(\bar{x}) - \bar{f}(\bar{y}) \right| \leq m \left| \bar{x} - \bar{y} \right| \tag{11}$$

for $\bar{x}$ and $\bar{y}$ in R, then the sequence $\bar{x}^{(0)}$, $\bar{x}^{(1)}$, $\bar{x}^{(2)}$, . . . converges to a solution of the initial-value problem (2), (3).

PROOF: First we note from (10) for $k = 0$ that

$$\bar{x}^{(1)} - \bar{x}^{(0)} = \bar{c} - \bar{x}^{(0)} + \int_0^t \bar{f}(\bar{x}^{(0)}(s))\, ds$$

and so by (9),

$$\left|\,\bar{x}^{(1)} - \bar{x}^{(0)}\,\right| \le \left|\,\bar{c} - \bar{x}^{(0)}\,\right| + \int_0^t \left|\,\bar{f}(\bar{x}^{(0)}(s))\,\right| ds \le a + Kt \le 2a$$

$$(12)$$

for $0 \le t \le b$.   From (10) for $k = 1$, we obtain

$$\left|\,\bar{x}^{(2)} - \bar{x}^{(1)}\,\right| = \left|\,\int_0^t \bar{f}(\bar{x}^{(1)}(s))\, ds - \int_0^t \bar{f}(\bar{x}^{(0)}(s))\, ds\,\right|$$

$$\le \int_0^t \left|\,\bar{f}(\bar{x}^{(1)}(s)) - \bar{f}(\bar{x}^{(0)}(s))\,\right| ds$$

Since each of $\bar{x}^{(1)}(s)$ and $\bar{x}^{(0)}(s)$ is in R for $0 \le s \le b$, we have, using (11),

$$\left|\,\bar{x}^{(2)} - \bar{x}^{(1)}\,\right| \le \int_0^t m \left|\,\bar{x}^{(1)}(s) - \bar{x}^{(0)}(s)\,\right| ds$$

which, together with (12), yields

$$\left|\,\bar{x}^{(2)} - \bar{x}^{(1)}\,\right| \le m \int_0^t 2a\, ds = 2a(mt) \qquad (13)$$

for $0 \le t \le b$.   More generally, we have for $k > 0$

$$\left|\,\bar{x}^{(k+1)} - \bar{x}^{(k)}\,\right| = \left|\,\int_0^t [\bar{f}(\bar{x}^{(k)}(s)) - \bar{f}(\bar{x}^{(k-1)}(s))]\, ds\,\right|$$

$$\le \int_0^t \left|\,\bar{f}(\bar{x}^{(k)}(s)) - \bar{f}(\bar{x}^{(k-1)}(s))\,\right| ds$$

and since each of $\bar{x}^{(k)}(s)$ and $\bar{x}^{(k-1)}(s)$ is in R for $0 \le s \le b$, we have, using (11),

$$\left|\,\bar{x}^{(k+1)} - \bar{x}^{(k)}\,\right| \le m \int_0^t \left|\,\bar{x}^{(k)}(s) - \bar{x}^{(k-1)}(s)\,\right| ds \qquad (14)$$

for $0 \le t \le b$.   Now if

$$\left|\,\bar{x}^{(k)}(t) - \bar{x}^{(k-1)}(t)\,\right| \le 2a \frac{(mt)^{k-1}}{(k-1)!} \qquad (15)$$

for $0 \le t \le b$, then

$$\left|\,\bar{x}^{(k+1)} - \bar{x}^{(k)}\,\right| \le \int_0^t \frac{2am^k}{(k-1)!} s^{k-1}\, ds = 2a \frac{(mt)^k}{k!} \qquad (16)$$

for $0 \leq t \leq b$. But (16) is (15) with k replaced by $k + 1$ and since (13) is (15) for $k = 2$, we have (by mathematical induction) that (15) holds for *all* $k \geq 2$. Now for $k \geq 2, l > 0$ we have

$$\left| \bar{x}^{(k+l)} - \bar{x}^{(k)} \right|$$
$$= \left| \bar{x}^{(k+l)} - \bar{x}^{(k+l-1)} + \bar{x}^{(k+l-1)} - \cdots + \bar{x}^{(k+1)} - \bar{x}^{(k)} \right|$$
$$\leq \left| \bar{x}^{(k+l)} - \bar{x}^{(k+l-1)} \right| + \left| \bar{x}^{(k+l-1)} - \bar{x}^{(k+l-2)} \right| + \cdots$$
$$+ \left| \bar{x}^{(k+1)} - \bar{x}^{(k)} \right|$$

which, together with (15), implies that

$$\left| \bar{x}^{(k+l)} - \bar{x}^{(k)} \right| \leq 2a \left[\frac{(mt)^{k+l-1}}{(k+l-1)!} + \frac{(mt)^{k+l-2}}{(k+l-2)!} \right.$$
$$\left. + \cdots + \frac{(mt)^k}{k!} \right] \quad (17)$$
$$< 2a \left[\frac{(mb)^k}{k!} + \frac{(mb)^{k+1}}{(k+1)!} + \cdots \right]$$
$$< 2ae^{mb} \frac{(mb)^k}{k!}$$

for $0 \leq t \leq b$. But $\lim\limits_{k \to \infty} \left[2ae^{mb} \frac{(mb)^k}{k!} \right] = 0$, and so (17), in turn, implies that the sequence $\bar{x}^{(0)}, \bar{x}^{(1)}, \bar{x}^{(2)}, \ldots$ converges (in the sense of Cauchy) uniformly on $0 \leq t \leq b$. Clearly then, if we denote by $\bar{x}$ the limit, we have from (10)

$$\bar{x} = \lim_{k \to \infty} \bar{x}^{(k+1)} = \bar{c} + \lim_{k \to \infty} \int_0^t \bar{f}(\bar{x}^{(k)}(s)) \, ds$$
$$= \bar{c} + \int_0^t \bar{f}(\lim_{k \to \infty} \bar{x}^{(k)}(s)) \, ds$$
$$= \bar{c} + \int_0^t \bar{f}(\bar{x}(s)) \, ds \quad (18)$$

since the convergence is uniform for $0 \leq t \leq b$, and $\bar{f}$ is uniformly continuous. Thus the theorem is proved.

EXERCISES

1. Demonstrate the equivalence of the initial-value problem (2), (3), and the integral equation problem (4).

2. Justify (18) by establishing the inequalities

$$\left| \bar{x}^{(k+1)} - \bar{c} - \int_0^t \bar{f}(\bar{x}(s))\, ds \right| \leq \int_0^t \left| \bar{f}(\bar{x}^{(k)}(s)) - \bar{f}(\bar{x}(s)) \right| ds$$

$$\leq m \int_0^t \left| \bar{x}^{(k)}(s) - \bar{x}(s) \right| ds$$

3. In the notation of Theorem 1, let $b_1 = \min\left(b, \dfrac{1}{m}\right)$ and $0 < b_2 < b_1$. Then show that

$$\left| \bar{x}^{(k+1)}(t) - \bar{x}^{(k)}(t) \right| \leq \left(\frac{b_2}{b_1}\right) \max_{0 \leq s \leq b_2} \left| \bar{x}^{(k)}(s) - \bar{x}^{(k-1)}(s) \right|$$

for $0 \leq t \leq b_2$, and hence

$$\max_{0 \leq t \leq b_2} \left| \bar{x}^{(k+1)}(t) - \bar{x}^{(k)}(t) \right| \leq \left(\frac{b_2}{b_1}\right) \max_{0 \leq s \leq b_2} \left| \bar{x}^{(k)}(s) - \bar{x}^{(k-1)}(s) \right|$$

Explain why this implies that

$$\max_{0 \leq t \leq b_2} \left| \bar{x}^{(k+1)}(t) - \bar{x}^{(k)}(t) \right| \leq 2a \left(\frac{b_2}{b_1}\right)^k$$

and that the sequence $\bar{x}^{(0)}, \bar{x}^{(1)}, \bar{x}^{(2)}, \ldots$ converges uniformly for $0 \leq t \leq b_2$.

4. Show that the transformation T given in (6) is "continuous." Take as the domain of T the set S with the uniform topology, i.e., $\lim_{k \to \infty} \bar{x}^{(k)} = \bar{x}$ in S means $\lim_{k \to \infty} \bar{x}^{(k)}(t) = \bar{x}(t)$ uniformly for $0 \leq t \leq b$. Then T is *continuous* if $\lim_{k \to \infty} \bar{x}^{(k)} = \bar{x}$ implies $\lim_{k \to \infty} T(\bar{x}^{(k)}) = T(\bar{x})$.

5. Using Exercise 4, show that the limit of the sequence of successive approximations of Theorem 1 is a fixed point of T.

2. *The Uniqueness Theorem*

THEOREM 2

With the hypotheses of Theorem 1, the solution of the initial-value problem is unique.

PROOF: Suppose each of $\bar{x}$ and $\bar{z}$ is a solution of the initial-value problem. Then

$$\bar{x} = \bar{c} + \int_0^t \bar{f}(\bar{x}(s))\, ds \qquad \text{and} \qquad \bar{z} = \bar{c} + \int_0^t \bar{f}(\bar{z}(s))\, ds$$

for $0 \leq t \leq b$. Hence, subtracting, we have

$$\left| \bar{x}(t) - \bar{z}(t) \right| \leq \int_0^t \left| \bar{f}(\bar{x}(s)) - \bar{f}(\bar{z}(s)) \right| ds \leq m \int_0^t \left| x(s) - \bar{z}(s) \right| ds$$

where m is a Lipschitz constant. The scalar function

$$\Phi(t) = \left| \bar{x}(t) - \bar{z}(t) \right|$$

thus satisfies the inequality

$$\Phi(t) \leq c + m \int_0^t \Phi(s)\, ds \tag{19}$$

for every $c > 0$. On the other hand, (19) may be written

$$\frac{\Phi(t)}{c + m \int_0^t \Phi(s)\, ds} \leq 1$$

and, since $\Phi(t)$ is continuous,

$$\frac{1}{m} \frac{d}{dt} \left[\ln\left(c + m \int_0^t \Phi(s)\, ds \right) \right] \leq 1$$

This in turn implies that

$$\ln \left[\frac{c + m \int_0^t \Phi(s)\, ds}{c} \right] \leq mt$$

for $0 \leq t \leq b$, and hence

$$c + m \int_0^t \Phi(s)\, ds \leq ce^{mt} \tag{20}$$

Using (19) and (20), we have finally

$$\Phi(t) \leq ce^{mt} \tag{21}$$

for $0 \leq t \leq b$. Since (21) holds for every $c > 0$, necessarily $\Phi(t) = 0$ or, what is the same, $\bar{x}(t) = \bar{z}(t)$ for $0 \leq t \leq b$, which was to be proved.

3. *Continuity with Respect to Initial Values*

Because the treatment is so similar to that of the previous theorem, we now consider an important result which properly belongs in the following chapter.

THEOREM 3

With the hypotheses of Theorem 1, the solution of the initial-value problem is a continuous function of the initial vector.

PROOF: We obtain an upper bound to the growth of the difference $\left|\,\bar{x}(t) - \bar{y}(t)\,\right|$ as a function of t, for any two solutions $\bar{x}$ and $\bar{y}$ of the initial-value problem corresponding to two initial vectors $\bar{c}$ and $\bar{c}^*$, respectively. In fact, from

$$\bar{x}(t) = \bar{c} + \int_0^t \bar{f}(\bar{x}(s))\,ds \qquad \text{and} \qquad \bar{y}(t) = \bar{c}^* + \int_0^t \bar{f}(\bar{y}(s))\,ds$$

we have

$$\left|\,\bar{x}(t) - \bar{y}(t)\,\right| \le \left|\,\bar{c} - \bar{c}^*\,\right| + \int_0^t \left|\,\bar{f}(\bar{x}(s)) - \bar{f}(\bar{y}(s))\,\right| ds$$

or, upon using (11),

$$\left|\,\bar{x}(t) - \bar{y}(t)\,\right| \le \left|\,\bar{c} - \bar{c}^*\,\right| + m \int_0^t \left|\,\bar{x}(s) - \bar{y}(s)\,\right| ds \quad (22)$$

The inequality (22) is of the form (19) with $\Phi(t) = \left|\,\bar{x}(t) - \bar{y}(t)\,\right|$ and $c = \left|\,\bar{c} - \bar{c}^*\,\right|$. Thus by (21) we have

$$\left|\,\bar{x}(t) - \bar{y}(t)\,\right| \le \left|\,\bar{c} - \bar{c}^*\,\right| e^{mt} \tag{23}$$

Clearly, (23) implies that the solution of the initial-value problem is a continuous function of the initial vector.

EXERCISES

1. That (19) implies (21) is a version of Gronwall's lemma.[1] Substantiate the following version. If Φ and ψ are nonnegative

[1] T. H. Gronwall, Note on the Derivatives with Respect to a Parameter of the Solutions of a System of Differential Equations, *Ann. Math.*, sec. 2, vol. 20, pp. 292–296, 1919.

continuous functions satisfying the inequality $\Phi(t) \leq c + \int_0^t \Phi(s)\,\psi(s)\,ds$ for $t \geq 0$, then it follows that

$$\Phi(t) \leq c \exp\left(\int_0^t \psi(s)\,ds\right)$$

for $t \geq 0$. Rephrase the lemma in terms of an initial-value problem for a differential inequality.

2. Prove Theorem 2 by showing that for any sequence of successive approximations (10) and for any $\bar{z}$ satisfying

$$\bar{z} = \bar{c} + \int_0^t \bar{f}(\bar{z}(s))\,ds$$

we have $\lim_{k \to \infty} |\bar{x}^{(k)}(t) - \bar{z}(t)| = 0$, uniformly for $0 \leq t \leq b$.

3. Reformulate Theorem 3 abstractly in terms of the function space S with the uniform topology.

4. Consider the transformation (6) as a mapping T from a product space consisting of pairs of constant vectors $\bar{c}$ and function vectors $\bar{x}(t)$. Reformulate Theorems 1 and 2 abstractly in terms of T and the natural projections of the product space into the two coordinate spaces. Illustrate by a sketch. If the family of solutions of (4) is interpreted as a "curve," $\bar{x}$ versus $\bar{c}$ in the product space, what property of this curve is prescribed by Theorem 3?

4. The Cauchy-Peano Existence Theorem

The existence of a solution to the initial-value problem (2), (3) may be established without recourse to the Lipschitz property. However, if the right-hand member of (2) is not a Lipschitz function, in general more than one solution may exist. For example, the initial-value problem $dx/dt = x^{1/2}$, $x(0) = 0$ admits the trivial solution in addition to $x = t^2/4$. In the proof of Theorem 1, we define a sequence $\bar{x}^{(0)}, \bar{x}^{(1)}, \bar{x}^{(2)}, \ldots$ of successive approximations by an iterative procedure. Upon imposing the Lipschitz condition (11) we are able to show that the sequence converges uniformly for $0 \leq t \leq b$. From this it follows that the limit of the sequence is a solution of the initial-value problem. When the Lipschitz condition is dropped, we can no longer guarantee that the total sequence converges, but by recourse to a fundamental theorem from real variable theory, we could assert that a properly chosen subsequence of the iterates is uniformly convergent on $0 \leq t \leq b$. The theorem of Ascoli states that if $\bar{x}^{(0)}$, $\bar{x}^{(1)}$, $\bar{x}^{(2)}$, . . . is an infinite sequence of uniformly bounded equicontinuous functions on $0 \leq t \leq b$, $b > 0$, then there exists a subsequence which converges uniformly on $0 \leq t \leq b$. In the present case, each of the $\bar{x}^{(k)}$ satisfies:

$a.$ $\left| \bar{x}^{(k)}(t) \right| \leq \left| \bar{c} \right| + a$ for $0 \leq t \leq b$.

and

$b.$ $\left| \bar{x}^{(k)}(t_2) - \bar{x}^{(k)}(t_1) \right| \leq K|t_2 - t_1|$ for $0 \leq t_1 \leq b$, $0 \leq t_2 \leq b$.

The first asserts that the sequence is uniformly bounded, and the second, that the family of functions is equicontinuous (K is independent of t_1, t_2, and k). Thus by the Ascoli theorem, a subsequence is uniformly convergent for $0 \leq t \leq b$.[1] However, the limit need not be a solution of the initial-value problem (2), (3). Thus we are forced to consider a sequence of approximations which are inherently more closely tied to the integral equation. We consider a proof of the Cauchy-Peano existence theorem (Theorem 4) using a discrete analogue of (2). As might be anticipated, the Ascoli theorem plays a central role, and our principal objective will be to construct a uniformly bounded sequence of equicontinuous approximations to a solution of (2). In this instance, however, the approximations are supplied by difference equations rather than by an iterative procedure. A uniform limit of these approximations *will* furnish a solution of the initial-value problem (2), (3). For the sake of simplicity, we shall treat only the scalar case, $n = 1$.

[1] See Exercise 12 for a proof based directly upon properties a and b.

Consider the scalar equation

$$\frac{dx}{dt} = f(x) \tag{24}$$

and the initial condition

$$x(0) = c \tag{25}$$

where $f(x)$ is continuous and bounded for $|x - c| \le a$. Let $|f(x)| \le K$ for $|x - c| \le a$, and let $b = a/K$. For $k = 1$, $2, \ldots$, let $t_k = b/k$, so that $0, t_k, 2t_k, \ldots, (k-1)t_k, kt_k = b$ subdivides the interval $0 \le t \le b$ into k equal parts. For k fixed, we solve the explicit-difference equation [analogue of (24)]

$$\frac{x((j+1)t_k) - x(jt_k)}{(j+1)t_k - jt_k} = f(x(jt_k)) \qquad j = 0, 1, \ldots, k-1 \tag{26}$$

subject to the initial condition $x(0) = c$. Denote by $x_j{}^k$ the value of x for $t = jt_k$. Then (26) is equivalent to

$$x_{j+1}^k = x_j{}^k + t_k f(x_j{}^k) \qquad j = 0, 1, \ldots, k-1 \tag{27}$$

with $x_0{}^k = c$. Let $x^k(t)$ be the polygon connecting the points $(jt_k, x_j{}^k)$ for $j = 0, 1, \ldots, k$ (see Fig. 1). We wish to show that

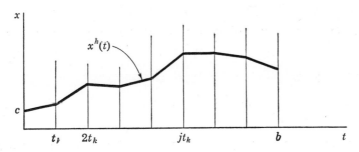

Figure 1

the family of polygons $x^k(t)$ consists of uniformly bounded equicontinuous functions. The arguments are analogous to the continuous case. In fact,

$$|x_j{}^k - c| \le (b) \max_{|x-c| \le a} |f(x)| \le bK = a \qquad \text{for all } k \text{ and } j$$

This follows easily from (27) and the definition of the quantity b. On the other hand, if $0 \leq t \leq b$ and $0 \leq \tau \leq b$, the slope of the secant between t and τ (see Fig. 2) does not exceed $K \geq \max |f(x)|$ for any polygon. That is, $|x^k(\tau) - x^k(t)| \leq K|\tau - t|$, where K is independent of τ, t, and k. Thus by the Ascoli theorem, there exists a subsequence x^{ν_1}, x^{ν_2}, ... of the polygons which is uniformly convergent to, say, x on $0 \leq t \leq b$.

In contrast to the continuous case where it was shown that the uniform limit was a solution of an equivalent integral equation,

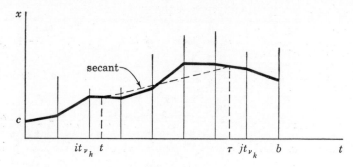

Figure 2

let us here show directly that the uniform limit is a solution of the differential equation (24). With reference to Fig. 2, if $it_{\nu_k} \leq t$ and $\tau \leq jt_{\nu_k}$, we have

$$\min_{i \leq r \leq j} f(x_r{}^{\nu_k}) \leq \frac{x^{\nu_k}(\tau) - x^{\nu_k}(t)}{\tau - t} \leq \max_{i \leq r \leq j} f(x_r{}^{\nu_k}) \qquad (28)$$

This states that the slope of the secant between t and τ lies between the minimum and the maximum of the slopes of the segments of the ν_kth polygon between it_{ν_k} and jt_{ν_k}. On the other hand,

$$\min_{it_{\nu_k} \leq s \leq jt_{\nu_k}} f(x^{\nu_k}(s)) \leq \min_{i \leq r \leq j} f(x_r{}^{\nu_k})$$

and

$$\max_{i \leq r \leq j} f(x_r{}^{\nu_k}) \leq \max_{it_{\nu_k} \leq s \leq jt_{\nu_k}} f(x^{\nu_k}(s))$$

since the continuous variables in each case cover larger sets than the discrete variables.   Hence (28) implies that

$$\min_{it_{\nu_k} \le s \le jt_{\nu_k}} f(x^{\nu_k}(s)) \le \frac{x^{\nu_k}(\tau) - x^{\nu_k}(t)}{\tau - t} \le \max_{it_{\nu_k} \le s \le jt_{\nu_k}} f(x^{\nu_k}(s)) \quad (29)$$

If we now assume (as implied in Fig. 2) that it_{ν_k} is always the nearest ν_k-subdivision point to the left of t and jt_{ν_k} is always the nearest ν_k-subdivision point to the right of τ, then with $k \to \infty$ (i.e., $\nu_k \to \infty$), $it_{\nu_k} \to t$ and $jt_{\nu_k} \to \tau$.   At the same time, $x^{\nu_k}(t) \to x(t)$, $x^{\nu_k}(\tau) \to x(\tau)$, and $x^{\nu_k}(s) \to x(s)$, the latter uniformly for $0 \le s \le b$.   Thus since $f(x)$ is uniformly continuous, (29) implies that

$$\min_{t \le s \le \tau} f(x(s)) \le \frac{x(\tau) - x(t)}{\tau - t} \le \max_{t \le s \le \tau} f(x(s)) \quad (30)$$

From (30) in turn, we have

$$\lim_{\tau \to t} \frac{x(\tau) - x(t)}{\tau - t} = f(x(t)) \quad (31)$$

and this shows that $x(t)$ is both differentiable and satisfies (24). Of course, $x(0) = c$ since $x^{\nu_k}(0) = c$ for each k.   This completes the proof for the scalar case of the fundamental theorem 4.

THEOREM 4

If the right-hand member in (2) is continuous in a neighborhood of an initial point $\bar{c}$, then there exists a solution of the initial-value problem (2), (3).

EXERCISES

1. If $0 < \alpha < 1$, show that the initial-value problem

$$\frac{dx}{dt} = |x|^\alpha$$

$x(0) = 0$ has infinitely many solutions.   In fact, show that for any $c > 0$, there exists a solution satisfying $x = 0$ for $0 \le t \le c$ and $x \ne 0$ for $c < t$.   Note that the right-hand member is continuous, but is not a Lipschitz function near the initial point.

2. Generalize Exercise 1 by considering the initial value problem $dx/dt = f(x)$, $x(0) = 0$ with $f(x)$ continuous and positive for $x \neq 0$ and $f(0) = 0$. Show that if $dx/f(x)$ has an integrable singularity at $x = 0$, then the initial-value problem has infinitely many solutions.

3. Show that the initial-value problem $dx/dt = x \ln |x|$, $x(0) = 0$ admits only the trivial solution by considering the integrability of $dx/x \ln x$ near $x = 0$. On the other hand, show that the right-hand member is continuous at $x = 0$ (define the value of the right-hand member appropriately for $x = 0$) but is not a Lipschitz function near $x = 0$. Explain the significance of such an example.

4. Generalize Exercise 3 by showing that the initial-value problem in Exercise 2 has a unique solution if $dx/f(x)$ is not integrable at $x = 0$.

5. Extend Exercise 4 by considering the vector equation $d\bar{x}/dt = \bar{f}(\bar{x})$, for which the right-hand member satisfies the inequality $|\bar{f}(\bar{x}) - \bar{f}(\bar{y})| \leq f(|\bar{x} - \bar{y}|)$, with $f(x)$ continuous and nondecreasing, $f(0) = 0$ and $\int dx/f(x)$ divergent at $x = 0$. Explain how this includes the Cauchy-Lipschitz case, Theorem 2.

6. Verify the properties a and b of this section.

7. Show by example that the limit of a uniformly convergent subsequence of successive approximations need not satisfy the integral equation (4).

8. Explain in detail why (30) follows from (29).

9. Show that the limit of the uniformly convergent subsequence of polygonal approximations x^{r_k} satisfies the integral equation

$$x(t) = c + \int_0^t f(x(s)) \, ds.$$

10. Prove the Cauchy-Peano existence theorem for two-dimensional systems ($n = 2$) using approximations supplied by difference equations.

11. Rephrase Theorem 4 for nonautonomous systems.

12. Let each of the vector functions $\bar{x}_k(t)$, $k = 1, 2, \ldots$ be defined on $0 \leq t \leq b$ with $b > 0$. If K and L are positive constants satisfying

$$\left|\,\bar{x}_k(t)\,\right| \leq L \quad \text{and} \quad \left|\,\bar{x}_k(t_1) - \bar{x}_k(t_2)\,\right| \leq K|t_1 - t_2|$$

for all $k = 1, 2, \ldots$ and t, t_1, t_2 in the interval $0 \leq t \leq b$, then one may show that a subsequence of the $\bar{x}_k(t)$ is uniformly convergent on $0 \leq t \leq b$. In fact, if we first arrange the rational points of the interval $0 \leq t \leq b$ in a sequence $r_1, r_2, \ldots$, then from among the $\bar{x}_k(t)$ we may choose a subsequence $\bar{x}_{1j}(t)$, $j = 1, 2, \ldots$ so that $\lim_{j \to \infty} \bar{x}_{1j}(r_1)$ exists. Why is this possible? Then from among the $\bar{x}_{1j}(t)$, we may choose a subsequence $\bar{x}_{2j}(t)$, $j = 1, 2, \ldots$ so that $\lim_{j \to \infty} \bar{x}_{2j}(r_2)$ exists. Continuing, we choose subsequences in such a fashion that for each positive integer i, $\bar{x}_{ij}(t)$, $j = 1, 2, \ldots$ is a subsequence of the sequence $\bar{x}_{(i-1)j}(t)$, $j = 1, 2, \ldots$ and $\lim_{j \to \infty} \bar{x}_{ij}(r_i)$ exists. Show that this is indeed possible and that the subsequence $\bar{x}_{11}(t)$, $\bar{x}_{22}(t)$, $\ldots$ converges at each of the rational points of the interval. This is called the *diagonalization process* since the subsequence $\bar{x}_{ii}(t)$, $i = 1, 2, \ldots$ consists of the diagonal elements of the infinite matrix $(\bar{x}_{ij})$. Now if $\epsilon > 0$, one may divide the interval $0 \leq t \leq b$ into a finite number of subintervals each of length less than $\epsilon/3K$. Interior to each subinterval we choose a representative rational point r_k. Show that there exists a positive integer N such that $i \geq N$ and $j \geq N$ imply $\left|\,\bar{x}_{ii}(r_k) - \bar{x}_{jj}(r_k)\,\right| \leq \epsilon/3$ for each of these finitely many r_k. Thus for such i and j and for any t in $0 \leq t \leq b$, one has (why?)

$$\left|\,\bar{x}_{ii}(t) - \bar{x}_{jj}(t)\,\right| \leq \left|\,\bar{x}_{ii}(t) - \bar{x}_{ii}(r_k)\,\right| + \left|\,\bar{x}_{ii}(r_k) - \bar{x}_{jj}(r_k)\,\right|$$
$$+ \left|\,\bar{x}_{jj}(r_k) - \bar{x}_{jj}(t)\,\right| < \epsilon$$

with an appropriate choice (which?) of r_k. Explain how this proves the desired result.

13. Prove the theorem of Ascoli.

5. *An Existence and Uniqueness Theorem for Nonautonomous Systems*

There are many circumstances in which one needs to relax the hypotheses in Theorems 1, 2, and 4 with reference to the independent variable. In such a situation, one is concerned with a nonautonomous system

$$\frac{d\bar{x}}{dt} = \bar{f}(\bar{x}, t) \tag{32}$$

for which the right-hand member $\bar{f}(\bar{x}, t)$ as a function of t is considerably less regular than it is as a function of the dependent variables $x_1, x_2, \ldots, x_n$. The artifice of introducing an auxiliary variable $x_{n+1} = t$ and thereby rendering (32) autonomous, throws (32) into the realms of Theorems 1, 2, and 4 but unnecessarily restricts the applications. Very often the difficulties are encountered only for an instant, say $t = 0$, where the right-hand member in (32) diverges. If the corresponding singularity is integrable, however, the difficulties are merely superficial. A companion to Theorems 1 and 2 for the nonautonomous case is Theorem 5.

THEOREM 5

Let $\bar{f}(\bar{x}, t)$ be continuous in $\bar{x}$ and t for $|\bar{x} - \bar{c}| \leq a$ and $0 < t \leq b$. Further, let $\bar{f}(\bar{x}, t)$ satisfy the inequalities

$$|\bar{f}(\bar{x}, t)| \leq K(t) \tag{33}$$

and
$$|\bar{f}(\bar{x}, t) - \bar{f}(\bar{y}, t)| \leq m(t)|\bar{x} - \bar{y}| \tag{34}$$

for $|\bar{x} - \bar{c}| \leq a$ and $0 < t \leq b$, where each of $K(t)$ and $m(t)$ is integrable on $0 \leq t \leq b$. Let $b_1(0 < b_1 \leq b)$ be chosen so that

$$\int_0^{b_1} K(t)\, dt \leq a, \tag{35}$$

and
$$\int_0^{b_1} m(t)\, dt = \delta < 1 \tag{36}$$

Then if $\bar{x}^{(0)}(t)$ is continuous, with $\left| \bar{x}^{(0)} - \bar{c} \right| \leq a$ for $0 < t \leq b_1$, and

$$\bar{x}^{(k+1)}(t) = \bar{c} + \int_0^t \bar{f}(\bar{x}^{(k)}(s), s) \, ds \tag{37}$$

for $k = 0, 1, 2, \ldots$, the sequence $\bar{x}^{(0)}$, $\bar{x}^{(1)}$, ... converges uniformly on $0 \leq t \leq b_1$ to a unique solution $\bar{x}$ of the integral equation

$$\bar{x}(t) = \bar{c} + \int_0^t \bar{f}(\bar{x}(s), s) \, ds \tag{38}$$

PROOF: From (33), (35), and (37) it follows that

$$\left| \bar{x}^{(k+1)} - \bar{c} \right| \leq \int_0^t \left| \bar{f}(\bar{x}^{(k)}(s), s) \right| ds \leq \int_0^t K(t) \, dt \leq a$$

for $0 \leq t \leq b_1$, provided

$$\left| \bar{x}^{(k)} - \bar{c} \right| \leq a \tag{39}$$

for $0 \leq t \leq b_1$. Thus since (39) is true for $k = 0$, it is true for all k. Now

$$\left| \bar{x}^{(1)} - \bar{x}^{(0)} \right| \leq \left| \bar{c} - \bar{x}^{(0)} \right| + \int_0^t \left| \bar{f}(\bar{x}^{(0)}(s), s) \right| ds \leq a$$
$$+ \int_0^t K(t) \, dt \leq 2a$$

for $0 \leq t \leq b_1$, and so

$$\left| \bar{x}^{(2)} - \bar{x}^{(1)} \right| \leq \int_0^t \left| \bar{f}(\bar{x}^{(1)}(s), s) - \bar{f}(\bar{x}^{(0)}(s), s) \right| ds$$
$$\leq \int_0^t m(s) \left| \bar{x}^{(1)}(s) - \bar{x}^{(0)}(s) \right| ds \leq 2a \int_0^t m(s) \, ds \leq 2a\delta$$

More generally,

$$\left| \bar{x}^{(k+1)} - \bar{x}^{(k)} \right| \leq \int_0^t m(s) \left| \bar{x}^{(k)}(s) - \bar{x}^{(k-1)}(s) \right| ds \tag{40}$$

and so if

$$\left| \bar{x}^{(k)} - \bar{x}^{(k-1)} \right| \leq 2a\delta^{k-1} \tag{41}$$

for $0 \leq t \leq b_1$, then

$$\left| \bar{x}^{(k+1)} - \bar{x}^{(k)} \right| \leq 2a\delta^{k-1} \int_0^t m(s) \, ds \leq 2a\delta^k$$

for $0 \leq t \leq b_1$, which is (41) with k replaced by $k + 1$. Thus since (41) is true for $k = 1$, it is true for all $k \geq 1$.

Now for $k \geq 1$ and $l > 0$, we have

$$\left| \bar{x}^{(k+l)} - \bar{x}^{(k)} \right| \leq \left| \bar{x}^{(k+l)} - \bar{x}^{(k+l-1)} \right| + \left| \bar{x}^{(k+l-1)} - \bar{x}^{(k+l-2)} \right|$$
$$+ \cdots + \left| \bar{x}^{(k+1)} - \bar{x}^{(k)} \right| \leq 2a(\delta^{k+l-1} + \delta^{k+l-2} + \cdots + \delta^{k})$$
$$= 2a\delta^{k}(1 + \delta + \delta^2 + \cdots + \delta^{l-1}) \quad (42)$$

Owing to (36), the last factor on the right in (42) is bounded by a convergent geometric series whose sum is $1/(1 - \delta)$. Hence

$$\left| \bar{x}^{(k+l)} - \bar{x}^{(k)} \right| \leq 2a\delta^k/(1 - \delta)$$

for $0 \leq t \leq b_1$. But $\lim\limits_{k \to \infty} 2a\delta^k/(1 - \delta) = 0$, and so the sequence $\bar{x}^{(0)}, \bar{x}^{(1)}, \bar{x}^{(2)}, \ldots$ converges uniformly on $0 \leq t \leq b_1$. It is not difficult to show that the uniform limit, say $\bar{x}$, satisfies the integral equation (38). In fact, using (34) and (36) we have for $k = 1, 2, \ldots,$

$$\left| \int_0^t \bar{f}(\bar{x}(s), s) \, ds - \int_0^t \bar{f}(\bar{x}^{(k)}(s), s) \, ds \right|$$
$$\leq \int_0^t m(s) \left| \bar{x}(s) - \bar{x}^{(k)}(s) \right| ds \leq \delta \max_{0 \leq s \leq b_1} \left| \bar{x}(s) - \bar{x}^{(k)}(s) \right|$$

for $0 \leq t \leq b_1$, and so

$$\int_0^t \bar{f}(\bar{x}(s), s) \, ds = \lim_{k \to \infty} \int_0^t \bar{f}(\bar{x}^{(k)}(s), s) \, ds$$
$$= \lim_{k \to \infty} [\bar{x}^{(k+1)}(t) - \bar{c}] = \bar{x}(t) - \bar{c}$$

for $0 \leq t \leq b_1$.

Uniqueness follows immediately from (36). In fact, if

$$\bar{x} = \bar{c} + \int_0^t \bar{f}(\bar{x}(s), s) \, ds \qquad \text{and} \qquad \bar{z} = \bar{c} + \int_0^t \bar{f}(\bar{z}(s), s) \, ds$$

then $$\left| \bar{x} - \bar{z} \right| \leq \int_0^t m(s) \left| \bar{x}(s) - \bar{z}(s) \right| ds \qquad (43)$$

Thus if Δ is the maximum of $\left| \bar{x}(t) - \bar{z}(t) \right|$ for $0 \leq t \leq b_1$, then

$$\left| \bar{x}(t) - \bar{z}(t) \right| \leq \Delta \int_0^t m(s) \, ds \leq \Delta\delta$$

holds for $0 \leq t \leq b_1$. Hence $\Delta \leq \Delta\delta$ which, in view of (36), is impossible unless $\Delta = 0$.

EXERCISES

1. Solve the initial-value problem $dx/dt = x^2/|t|^{\frac{1}{2}}$, $x(0) = 1$. Illustrate each detail of the statement and the proof of Theorem 5 for this special case. Consider the initial condition $x(0) = 0$.

2. Show that the initial-value problem $dx/dt = 2x/t^3$, $x(0) = 0$ has infinitely many solutions. What hypotheses of Theorem 5 are not fulfilled in this case?

3. Show that the initial-value problem $dx/dt = |tx|^{\frac{1}{2}}$, $x(0) = 0$ has infinitely many solutions. What hypotheses of Theorem 5 are not fulfilled in this case?

4. Show that the initial-value problem $dx/dt = t/x^2$, $x(0) = 0$ has a unique solution. Explain the significance of such an example.

5. Generalize Theorem 5 to cover initial-value problems for arbitrary initial times t_0.

6. Using Gronwall's lemma and (43), prove that the solution $\bar{x}$ of (38) is unique. First extend Gronwall's lemma so that it applies to integrable (but not necessarily continuous) functions.

7. State and prove a companion to Theorem 3 for the non-autonomous system (38).

8. Prove Theorem 5 in the scalar case, $n - 1$, using a sequence of approximations supplied by difference equations.

Chapter 3

PROPERTIES OF SOLUTIONS

1. *Extension of the Trajectories*

The existence theorems of Chap. 2 concern a solution of the initial-value problem

$$\frac{d\bar{x}}{dt} = \bar{f}(\bar{x}) \qquad \bar{x}(0) = \bar{c} \tag{1}$$

for a certain interval $0 \leq t \leq b$ of the independent variable. If the right-hand member of the system is a Lipschitz function, then the solution is unique. As we proceed along a trajectory $\bar{x}(t)$ from its initial position $\bar{c}$, we encounter, for each t_1 satisfying $0 \leq t_1 \leq b$, points $\bar{x}(t_1)$ of n-space each of which may be considered an appropriate initial position for a new initial-value problem. In general, the t interval of existence increases indefinitely (i.e., the trajectory beginning at $\bar{c}$ exists for all $t \geq 0$) or we arrive, for $t = t_2$, at a position $\bar{c}^*$ such that the initial-value problem for $t \geq t_2$ with initial value $\bar{x}(t_2) = \bar{c}^*$ has no solution, or a component of the solution $\bar{x}(t)$ diverges as $t \to t_2$. For example, the initial-value problem $dx/dt = 1/(1 - x)$, $x(0) = 0$ leads to the trajectory $x = 1 - \sqrt{1 - 2t}$, which arrives at the point $x = 1$ for $t = \frac{1}{2}$ and cannot be continued for larger t, while the initial-value problem $dx/dt = x^2$, $x(0) = 1$ leads to the trajectory $x(t) = 1/(1 - t)$, which is mapped out completely with

60

$0 \le t < 1$. On the other hand, if $\lim_{t \to t_2} \bar{x}(t) = \bar{c}^*$ exists and $\bar{f}(\bar{x})$
is continuous near $\bar{c}^*$, then a solution satisfying $\bar{x}(t_2) = \bar{c}^*$ exists
on some interval $t_2 \le t \le t_2 + b_1$ with $b_1 > 0$ and the latter
becomes an extension of the former. Thus if one considers only
systems which are continuous throughout, then a solution $\bar{x}(t)$ of
an initial-value problem may be continued for all $t \ge 0$ or a component diverges as $t \to t_2$ with $0 < t_2 < \infty$.

In this chapter and subsequent chapters, the word *trajectory*
will generally connote the unique solution of an initial-value
problem (forward and/or backward) and all possible (unique)
extensions for $t \ge 0$ (and/or $t \le 0$). That portion of a trajectory
mapped out for $t \ge 0$ will be called a *positive half path*, and that
portion mapped out for $t \le 0$ will be called a *negative half path*.
If a positive half path approaches a finite point $\bar{c}$ as $t \to \infty$, then $\bar{c}$
is necessarily a singular point. In fact, if $\bar{f}(\bar{c}) \ne \bar{0}$, then $\|\bar{f}(\bar{x})\|$
is bounded away from zero for $\bar{x}$ near $\bar{c}$ and clearly if $\lim_{t \to \infty} \bar{x}(t) = \bar{c}$,
at least one component of the integral $\int_0^t \bar{f}(\bar{x}(s))\, ds$ diverges as
$t \to \infty$. But for a trajectory, the difference $\bar{x}(t) - \int_0^t \bar{f}(x(s))\, ds$
must be constant, which is impossible. On the other hand, if
each initial-value problem admits but one solution, then a positive half path can approach a singular point only for $t \to \infty$.
More generally, we define the positive *limit cycle* of a trajectory Γ
as the set of those points which are near Γ for $t \to \infty$. We shall
be more precise: Let $\bar{c}$ belong to Γ, and let $\gamma(\bar{c})$ denote the positive
half path associated with the initial value $\bar{c}$. Then the positive
limit cycle of Γ is the intersection of the sets of limit points of
$\gamma(\bar{c})$ as $\bar{c}$ varies on Γ. It is not difficult to show that a point $\bar{b}$
belongs to the positive limit cycle of Γ if and only if the trajectory
intersects each neighborhood of $\bar{b}$ for arbitrarily large values of t.
Typically, a limit cycle is a singular point, a closed trajectory
containing no singular points (corresponding to a periodic solution
of the differential equation), an empty set (i.e., nonexistent), or a

collection of singular points and connecting paths called *separatrices*. The first two, singular points and periodic solutions, are the limit cycles of principal concern in this book.

2. *Continuity Properties of Trajectories*

A solution $\bar{x}(t)$ of the initial-value problem (1) is continuous in t, i.e., the trajectory is an arc in $x_1, x_2, \ldots, x_n, t$ space. In fact, $\bar{x}(t)$ is continuously differentiable in t so that a nonsingular trajectory possesses a continuously turning tangent and its arc length is always defined. More generally, if the system possesses kth-order continuous derivatives, then a solution $\bar{x}(t)$ possesses a a $(k + 1)$st continuous derivative with respect to t.

According to Theorem 3 of Chap. 2, the solution $\bar{x}(t)$ of (1) is a continuous function of the initial vector $\bar{c}$. This follows also from the fact that $\bar{x}(t)$ is the uniform limit of a sequence of functions each of which is a continuous function of the initial vector $\bar{c}$. The latter argument employs a powerful tool which permits one to "transfer" a property of approximates to a solution. We shall have use of it again. But for the moment, let us generalize these important continuity properties.

By introducing auxiliary dependent variables, one shows that the solution $\bar{x}(t)$ of the initial-value problem is a continuous function of any parameters which affect the system in a continuous fashion. In particular, if $\bar{f}(\bar{x}) = \bar{f}_{\alpha_1, \alpha_2, \ldots, \alpha_j}(\bar{x})$ is a continuous vector function of the j scalar parameters $\alpha_1, \alpha_2, \ldots, \alpha_j$, then one adjoins to the vector $\bar{x}$, j components $x_{n+1}, x_{n+2}, \ldots, x_{n+j}$ and to the differential system (1), the j equations

$$\frac{dx_{n+1}}{dt} = 0$$

$$\frac{dx_{n+2}}{dt} = 0$$

$$\cdots \cdots$$

$$\frac{dx_{n+j}}{dt} = 0$$

subject to the initial conditions

$$x_{n+1}(0) = \alpha_1$$
$$x_{n+2}(0) = \alpha_2$$
$$\cdot \cdot \cdot \cdot \cdot \cdot \cdot$$
$$x_{n+j}(0) = \alpha_j$$

By this artifice, the parameters $\alpha_1, \alpha_2, \ldots, \alpha_j$ appear as "initial values" of the $(n + j)$th-order system. The enlarged system is certainly continuous if (1) is, and the extended solution thus becomes a continuous function of the parameters $\alpha_1, \alpha_2, \ldots, \alpha_j$. The application of the technique used in the proof of Theorem 3 of Chap. 2 yields an upper bound to quantitative effects of variations of these parameters.

EXAMPLE 1

Consider the linear second-order equation

$$\frac{d^2x}{dt^2} + k\frac{dx}{dt} + \omega^2 x = 0 \qquad (2)$$

where $k \geq 0$, $\omega^2 > 0$, and the initial values $x = x_0$, $dx/dt = y_0$ for $t = 0$. Introduce the components $x_1 = x$, $x_2 = dx/dt$, $x_3 = k$, and $x_4 = \omega^2$. Then (2) is equivalent to the system

$$\frac{dx_1}{dt} = x_2$$
$$\frac{dx_2}{dt} = -x_4 x_1 - x_3 x_2$$
$$\frac{dx_3}{dt} = 0 \qquad (3)$$
$$\frac{dx_4}{dt} = 0$$

where the initial values become $x_1 = x_0$, $x_2 = y_0$, $x_3 = k$, and $x_4 = \omega^2$. Now (3) satisfies the Lipschitz condition

$$\left| \bar{f}(\bar{x}) - \bar{f}(\bar{y}) \right| = |x_2 - y_2| + |x_4 x_1 + x_3 x_2 - y_4 y_1 - y_3 y_2|$$
$$= |x_2 - y_2| + |x_4(x_1 - y_1) + x_3(x_2 - y_2) \quad y_1(y_4 - x_4)$$
$$- y_2(y_3 - x_3)| \leq |x_4| |x_1 - y_1| + (1 + |x_3|)|x_2 - y_2|$$
$$+ |y_2| |x_3 - y_3| + |y_1| |x_4 - y_4| \leq m(|x_1 - y_1| + |x_2 - y_2|$$
$$+ |x_3 - y_3| + |x_4 - y_4|) = m \left| \bar{x} - \bar{y} \right|$$

provided

$$|x_4| \leq m \qquad (1 + |x_3|) \leq m \qquad |y_2| \leq m \qquad \text{and} \qquad |y_1| \leq m \qquad (4)$$

If we define "neighboring" initial values $y_1 = x_0$, $y_2 = y_0$, $y_3 = k_*$, and $y_4 = \omega_*^2$ for a neighboring solution $y = y_1$ and denote by

$$\bar{y} = \begin{pmatrix} y_1 \\ y_2 \\ y_3 \\ y_4 \end{pmatrix} = \begin{pmatrix} y \\ \dfrac{dy}{dt} \\ k_* \\ \omega_*^2 \end{pmatrix}$$

the corresponding vector solution, then from Eq. (23) of Chap. 2 we have

$$
\begin{aligned}
|\bar{x} - \bar{y}| &= |x_1 - y_1| + |x_2 - y_2| + |x_3 - y_3| + |x_4 - y_4| \\
&= |x - y| + \left| \frac{dx}{dt} - \frac{dy}{dt} \right| + |k - k_*| + |\omega^2 - \omega_*^2| \\
&\leq |\bar{c} - \bar{c}^*| e^{mt} = (|k - k_*| + |\omega^2 - \omega_*^2|) e^{mt}
\end{aligned}
$$

or, what is the same,

$$|x - y| + \left| \frac{dx}{dt} - \frac{dy}{dt} \right| \leq (|k - k_*| + |\omega^2 - \omega_*^2|)(e^{mt} - 1)$$

The Lipschitz constant m may be any number which is an upper bound to the values of the quantities ω^2, $1 + |k|$, and $\sqrt{x_0^2 + (y_0/\omega)^2}$ to be considered. Such a choice for m guarantees that the four inequalities (4) will be satisfied along any two trajectories $\bar{x}$ and $\bar{y}$; i.e., the trajectories under consideration remain in a region for which the Lipschitz condition

$$|\bar{f}(\bar{x}) - \bar{f}(\bar{y})| \leq m |\bar{x} - \bar{y}|$$

is satisfied.

EXAMPLE 2

Consider the Mathieu equation

$$\frac{d^2x}{dt^2} + (\omega^2 + \epsilon \cos t)x = 0 \qquad (5)$$

where each of ω^2 and ϵ is a constant.   For $\epsilon = 0$, each solution of (5) is a simple harmonic function of frequency ω

$$y(t) = A \cos (\omega t + \Phi)$$

Consider the particular case $\Phi = 0$, corresponding to the initial conditions $y = A$, $dy/dt = 0$ for $t = 0$.   If $\epsilon \neq 0$, the situation may be quite complicated.   Nonetheless, let us define the components $x_1 = x$, $x_2 = dx/dt$, and $x_3 = \epsilon$ for a vector $\bar{x}$.   Then (5) is equivalent to the vector equation

$$\frac{d\bar{x}}{dt} = \bar{f}(\bar{x}) \tag{6}$$

where
$$\bar{f}(\bar{x}) = \begin{pmatrix} x_2 \\ -(\omega^2 + x_3 \cos t)x_1 \\ 0 \end{pmatrix}$$

Introduce the initial vector

$$\bar{c} = \begin{pmatrix} A \\ 0 \\ \epsilon \end{pmatrix}$$

and let $\bar{x}$ denote the solution of (6) satisfying $\bar{x} = \bar{c}$ for $t = 0$. Now the right-hand member in (6) satisfies the Lipschitz condition

$$\left| \bar{f}(\bar{x}) - \bar{f}(\bar{y}) \right|$$
$$= |x_2 - y_2| + |(\omega^2 + x_3 \cos t)x_1 - (\omega^2 + y_3 \cos t)y_1|$$
$$= |x_2 - y_2| + |\omega^2(x_1 - y_1) + \cos t(x_3 x_1 - y_3 y_1)|$$
$$\leq |x_2 - y_2| + \omega^2|x_1 - y_1| + |x_3| \, |x_1 - y_1| + |y_1| \, |x_3 - y_3|$$
$$\leq m \left| \bar{x} - \bar{y} \right|$$

provided

$$\omega^2 + |x_3| \leq m \qquad 1 \leq m \qquad \text{and} \qquad |y_1| \leq m \tag{7}$$

Let
$$\bar{c}^* = \begin{pmatrix} A \\ 0 \\ 0 \end{pmatrix}$$

be a "neighboring" initial vector, and let $\bar{y}$ denote the solution of (6) satisfying $\bar{y} = \bar{c}^*$ for $t = 0$.   Of course, this is merely the vectorial version of the simple harmonic solution $y(t)$. Using Eq. (23) of Chap. 2, we have

$$\| \bar{x} - \bar{y} \| = |x - y| + \left| \frac{dx}{dt} - \frac{dy}{dt} \right| + |\epsilon| \leq \| \bar{c} - \bar{c}^* \| e^{mt} = |\epsilon| e^{mt}$$

or, what is the same,

$$|x - y| + \left| \frac{dx}{dt} - \frac{dy}{dt} \right| \leq |\epsilon|(e^{mt} - 1) \tag{8}$$

According to (7), the Lipschitz constant m may be any number which is an upper bound to the number 1, to the values of $\omega^2 + |\epsilon|$ to be considered and to some quantity larger than $|A|$ which allows for a possible increase in amplitude. The inequality (8) will then be valid so long as the bound on the amplitude is not exceeded.

EXERCISES

1. Explain why it is that the solution of (1) possesses a $(k + 1)$st continuous derivative with respect to t if the right-hand member in (1) possesses kth-order continuous derivatives.

2. Explain the significance of the quantity $\sqrt{x_0{}^2 + (y_0/\omega)^2}$ introduced in Example 1. In what way should the results be modified if $k < 0$?

3. Show that (4) defines the "best possible" Lipschitz constant for (3).

4. Discuss the significance of the inequalities in (7) and the reference to amplitude variations following (8). Show how (8) itself may be employed to establish a bound on the amplitude variations.

5. Obtain an upper bound for the perturbations due to the nonlinear term in the satellite equation $d^2u/dt^2 + u = ku^2$.

6. Let $\bar{\alpha}$ denote a column vector with the components α_1, $\alpha_2, \ldots, \alpha_j$. Reformulate the results of this section for the

system $d\bar{x}/dt = \bar{f}(\bar{a},\bar{x})$, where $\bar{f}(\bar{a},\bar{x})$ is a Lipschitz function in each of the variables $\bar{a},\bar{x}$. Extend the technique used in the proof of Theorem 3 of Chap. 2 so as to include the effects on a solution of variations in the "parameter vector" $\bar{a}$.

3. *The Poincaré Expansion Theorem*

Here we apply the "transfer" technique as mentioned above to analytical or partly analytical systems. We are primarily concerned with the effects of parameter variations in the system. For the proofs of the results, we need not distinguish between those parameters which arise naturally in the form of initial values and those which are of a more parasitic nature. However, for practical emphasis we *do* distinguish the two types of parameters in the statements of the results.

It is clear from the equation

$$\bar{x}^{(k+1)}(t) = \bar{c} + \int_0^t \bar{f}(\bar{x}^{(k)}(s), s)\, ds \tag{9}$$

which [together with an initial choice $\bar{x}^{(0)}$] defines the successive approximations to a solution $\bar{x}(t)$ of an initial-value problem, that if $\bar{x}^{(0)}(t)$ is analytic in t (for example, if $\bar{x}^{(0)} = \bar{c}$) and if $\bar{f}(x,t)$ is analytic in the pair $(\bar{x},t)$, then $\bar{x}^k(t)$ is analytic in the pair $(\bar{c},t)$ for each $k = 1, 2, \ldots$. If $\bar{f}(\bar{x},t)$ is merely continuous in t but analytic in $\bar{x}$, then $\bar{x}^{(k)}(t)$ is analytic in $\bar{c}$ for each $k = 1, 2, \ldots$. Thus, since the sequence $\bar{x}^{(0)}, \bar{x}^{(1)}, \ldots$ converges uniformly in the pair $(\bar{c},t)$ to the solution of the initial-value problem, we have Theorem 1.

THEOREM 1

Consider the vector system

$$\frac{d\bar{x}}{dt} = \bar{f}_{\alpha_1, \alpha_2, \ldots, \alpha_j}(\bar{x}, t) \tag{10}$$

and the initial vector

$$\bar{c} = \begin{pmatrix} c_1 \\ c_2 \\ \cdots \\ c_n \end{pmatrix}$$

If the right-hand member in (10) is analytic in the $n + j + 1$ variables $x_1, x_2, \ldots, x_n, t, \alpha_1, \alpha_2, \ldots, \alpha_j$, then the solution $\bar{x}(t)$ of the initial-value problem $\bar{x}(0) = \bar{c}$ is analytic in the $n + j + 1$ variables, $c_1, c_2, \ldots, c_n, t, \alpha_1, \alpha_2, \ldots, \alpha_j$. If (10) is analytic in the $n + j$ variables, $x_1, x_2, \ldots, x_n, \alpha_1, \alpha_2, \ldots, \alpha_j$ and continuous in t, then the solution of the initial-value problem is analytic in the $n + j$ variables $c_1, c_2, \ldots, c_n, \alpha_1, \alpha_2, \ldots, \alpha_j$ and continuously differentiable in t.

This theorem has proved to be of great importance in the study of classical dynamical systems generally, and in celestial mechanics in particular.   It is the basis for almost all present-day perturbation theory.   One should note that this is a local theorem and concerns an initial-value problem for which, in general, the independent variable t is restricted to a *finite* interval.   Thus, for example, it is not directly applicable to problems concerning perturbations of periodic solutions.   There are, however, modifications of the expansion technique which apply to the latter.

Example 3

Consider the satellite equation

$$\frac{d^2u}{dt^2} + u = ku^2 \tag{11}$$

For $k = 0$, (11) possesses the periodic solution

$$u = A \cos t \tag{12}$$

where A is a constant.   Let us characterize (12) as the solution of the initial-value problem

$$u = A$$
$$\frac{du}{dt} = 0 \tag{13}$$

for $t = 0$.   If we apply the Poincaré expansion theorem, we know that for sufficiently small $|k|$, the solution of (11) satisfying (13) may be expressed as a power series in k.   Let

$$u = u_0 + u_1 k + u_2 k^2 + \cdots \tag{14}$$

be the expansion.   If we substitute (14) into (11) and equate coefficients of corresponding powers of k, we obtain a sequence of nested differential equations in $u_0, u_1, u_2 \ldots$.   The first three are

$$\frac{d^2 u_0}{dt^2} + u_0 = 0 \tag{15}$$

$$\frac{d^2 u_1}{dt^2} + u_1 = u_0{}^2 \tag{16}$$

$$\frac{d^2 u_2}{dt^2} + u_2 = 2u_0 u_1 \tag{17}$$

Clearly u_0 is given by (12), since (14) reduces to $u = u_0$ for $k = 0$.   Also, one sees that each of the u_i with $i \geq 1$ must satisfy homogeneous initial conditions.   Using (12), (16) becomes

$$\frac{d^2 u_1}{dt^2} + u_1 = \frac{A^2}{2} + \frac{A^2}{2} \cos 2t$$

and hence

$$u_1 = A^2(\tfrac{1}{2} - \tfrac{1}{3} \cos t - \tfrac{1}{6} \cos 2t) \tag{18}$$

where the initial conditions $u_1 = du_1/dt = 0$ for $t = 0$ have been imposed.   If (18), in turn, is used in (17), together with $u_0 = A \cos t$, we obtain

$$\frac{d^2 u_2}{dt^2} + u_2 = A^3(-\tfrac{1}{3} + \tfrac{5}{6} \cos t + \tfrac{1}{3} \cos 2t - \tfrac{1}{6} \cos 3t)$$

and so

$$u_2 = A^3(-\tfrac{1}{3} + \tfrac{29}{144} \cos t + \tfrac{5}{12}t \sin t + \tfrac{1}{9} \cos 2t + \tfrac{1}{48} \cos 3t)$$

We note that u_2 contains a resonance (secular) term, $t \sin t$. Because of the presence of this term and similar terms which occur in the expansion (14), there is no suggestion at all that the solution might be periodic, or if it is, what the fundamental period might be.   Yet for sufficiently small $|A|$, the solution *is* periodic. The secular terms arise because the frequencies of the expansion functions are incorrect.   The picture is cleared if we search for the true fundamental frequency at the same time we search for the periodic solution.

Following Poincaré,[1] we replace the expansion (14) with one of the form

$$u = y_0(\omega t) + y_1(\omega t)k + y_2(\omega t)k^2 + \cdots \tag{19}$$

where each of y_0, y_1 $\cdots$ is to be a periodic function of $\tau = \omega t$ of period 2π.   The quantity ω is introduced as the true fundamental frequency and will be an analytic function of k.   To determine ω, we introduce an expansion of the form

$$\omega = 1 + \omega_1 k + \omega_2 k^2 + \ldots \tag{20}$$

where the first term is unity since the fundamental frequency reduces to unity for $k = 0$.   The expansions (19) and (20) together represent a reorganization of the terms in (14) in such a way as to produce periodic coefficients for the power series in k, the coefficients themselves being power series in k.   In terms of the new variable $\tau = \omega t$, (11) becomes

$$\omega^2 \frac{d^2u}{d\tau^2} + u = ku^2$$

and the counterparts of (15), (16), and (17), using the expansions (19) and (20), become

$$\frac{d^2y_0}{d\tau^2} + y_0 = 0 \tag{21}$$

$$\frac{d^2y_1}{d\tau^2} + y_1 = y_0{}^2 - 2\omega_1 \frac{d^2y_0}{d\tau^2} \tag{22}$$

$$\frac{d^2y_2}{d\tau^2} + y_2 = 2y_0y_1 - (\omega_1{}^2 + 2\omega_2)\frac{d^2y_0}{d\tau^2} - 2\omega_1 \frac{d^2y_1}{d\tau^2} \tag{23}$$

[1] H. Poincaré, "Les Méthodes nouvelles de la mécanique céleste," vol. I, Gauthier-Villars, Paris, 1892.

Clearly, $y_0 = A \cos \tau$, and so (22) may be expressed

$$\frac{d^2y_1}{d\tau^2} + y_1 = \frac{A^2}{2} + 2\omega_1 A \cos \tau + \frac{A^2}{2} \cos 2\tau \qquad (24)$$

Unless $\omega_1 = 0$, every solution of (24) will contain a secular term. Thus we require that $\omega_1 = 0$ and eliminate the resonant term in (24) altogether. Then the appropriate solution becomes

$$y_1 = A^2(\tfrac{1}{2} - \tfrac{1}{3} \cos \tau - \tfrac{1}{6} \cos 2\tau)$$

which is periodic in τ with period 2π, as required. The third equation (23) becomes

$$\frac{d^2y_2}{d\tau^2} + y_2 = (\tfrac{5}{6}A^3 + 2A\omega_2) \cos \tau$$
$$- A^3(\tfrac{1}{3} + \tfrac{1}{3} \cos 2\tau + \tfrac{1}{6} \cos 3\tau) \quad (25)$$

Every solution of (25) will contain a secular term unless ω_2 is chosen so as to remove the first harmonic on the right. Hence, we require that $\omega_2 = -5A^2/12$ and the appropriate solution of (25) becomes

$$y_2 = A^3(-\tfrac{1}{3} + \tfrac{29}{144} \cos \tau + \tfrac{1}{9} \cos 2\tau + \tfrac{1}{48} \cos 3\tau)$$

In a similar manner, the remaining terms of the two expansions (19) and (20) are determined by the initial conditions and the requirement of periodicity. The technique of "casting out" the resonant terms is referred to as *Lindsted's procedure*.

It should be noted that the convergence of (19) will not be uniform in t for all t unless ω is *exact*, although it may be uniform in $\tau = \omega t$ for all τ. As an expansion in functions of t, (19) is referred to as an *asymptotic expansion*.

EXERCISES

1. Explain why it is that the successive approximations (9) converge uniformly in $(\bar{c}, t)$ to the unique solution of the analytic system (10). Refer to ranges of the variables.

2. By introducing a suitable artifice, use Theorem 1 to show

that the solution $\bar{x}(t)$ of (10) satisfying the initial condition $\bar{x}(t_0) = \bar{c}$ is analytic in the initial time t_0.

3. Discuss the power-series expansions in β of the solutions of the Duffing equation $d^2x/dt^2 + \omega^2 x + \beta x^3 = 0$. Obtain several terms of the expansion of the solution of an initial-value problem.

4. Obtain several terms of a power-series expansion in β of periodic solutions of the Duffing equation.

5. Reformulate the Poincaré expansion theorem (Theorem 1) for the system $d\bar{x}/dt = \bar{f}(\bar{a},\bar{x},t)$ where $\bar{a}$ is a column vector with components $\alpha_1, \alpha_2, \ldots, \alpha_j$.

4. Differentiability of Solutions

We have seen that the solution of an initial-value problem for a continuous (analytic) system varies continuously (analytically) with the initial values and the system parameters. It is natural, and for certain purposes also important, to consider the question of differentiability of the solutions. The "transfer" technique is not applicable here since differentiability of approximates will not necessarily be "passed on" to a uniform limit. It is necessary, therefore, to attack the problem directly. In view of the ease with which one deals with continuous and analytic systems in these matters, the proof of the following theorem is surprisingly difficult.

THEOREM 2

If $\bar{f}_{\alpha_1,\ldots,\alpha_j}(\bar{x},t)$ possesses kth-order continuous derivatives in the $n + j + 1$ variables $\alpha_1, \alpha_2, \ldots, \alpha_j, t, x_1, x_2, \ldots, x_n$, then the solution of the initial-value problem

$$\frac{d\bar{x}}{dt} = \bar{f}_{\alpha_1,\ldots,\alpha_j}(\bar{x},t)$$

$$\bar{x}(0) = \bar{c} = \begin{pmatrix} c_1 \\ c_2 \\ \cdots \\ c_n \end{pmatrix} \tag{26}$$

possesses kth-order continuous derivatives in the $n + j$ variables $\alpha_1, \alpha_2, \ldots, \alpha_j, c_1, c_2, \ldots, c_n$. Furthermore, the derivative $d\bar{x}/dt$ possesses kth-order continuous derivatives in the $n + j + 1$ variables $\alpha_1, \alpha_2, \ldots, \alpha_j, t, c_1, c_2, \ldots, c_n$.

PRELIMINARY REMARKS: In the proof of Theorem 2 one need not make a distinction between the two types of quantities α_i, c_i. For by the artifice where one appropriately enlarges the system, one can accommodate all the quantities as initial values. Alternatively, if one considers, in lieu of $\bar{x}$, the vector $\bar{y} = \bar{x} - \bar{c}$, then (26) is equivalent to the initial-value problem

$$\frac{d\bar{y}}{dt} = \bar{f}_{\alpha_1, \ldots, \alpha_j}(\bar{y} + \bar{c}, t)$$
$$\bar{y}(0) = \bar{0}$$

(27)

where now all the quantities α_i, c_i appear as system parameters and none as initial values. Since (27) is essentially of the same complexity as (26) (being of the same dimension and not of a higher dimension), there would apparently be some tactical advantage to employing the second of these two artifices. In point of fact, however, we shall give a proof of Theorem 2 only for the scalar case $n = 1$. The essential arguments for the general case are illustrated therein.[1]

In the proof we employ still another version of the Gronwall lemma.

LEMMA 1

If $\Phi(t)$, $\psi(t)$, and $\rho(t)$ are each nonnegative continuous functions for $t \geq 0$ and c is a positive constant such that $\Phi(t) \leq c + \int_0^t [\psi(s)\,\Phi(s) + \rho(s)]\,ds$ for $t \geq 0$, then

$$\Phi(t) \leq c \exp\left\{\int_0^t [\psi(s) + \rho(s)/c]\,ds\right\}$$

for $t \geq 0$.

---

[1] For a proof in the general case, see S. Lefschetz, "Differential Equations: Geometric Theory," pp. 40–43, Interscience Publishers, Inc., New York, 1957.

PROOF OF THEOREM 2 FOR $n = 1$: We consider the scalar equation

$$\frac{dx}{dt} = f(\alpha,x,t) \tag{28}$$

where $f(\alpha,x,t)$ is continuously differentiable in α, x, and t for α in some interval I, for x in some neighborhood of an initial value c, and $0 \leq t \leq b$, $b > 0$. Then for each α in I, (28) possesses a unique solution $x(\alpha,t)$ for $0 \leq t \leq b$, which satisfies $x(\alpha,0) = c$. We wish to show that $x(\alpha,t)$ is continuously differentiable with respect to α and that $dx\,(\alpha,t)/dt$ is continuously differentiable with respect to both α and t. Consider first the differentiability of $x(\alpha,t)$. Let α in I be fixed, and $h \neq 0$ be such that $\alpha + h$ is in I. Then we have for $0 \leq t \leq b$,

$$x(\alpha + h, t) = c + \int_0^t f(\alpha + h, x(\alpha + h, s), s)\, ds$$

and $$x(\alpha,t) = c + \int_0^t f(\alpha, x(\alpha,s), s)\, ds$$

The pertinent difference quotient may be expressed in the form

$$\frac{x(\alpha + h, t) - x(\alpha,t)}{h}$$
$$= \int_0^t \frac{f(\alpha + h, x(\alpha + h, s), s) - f(\alpha + h, x(\alpha,s), s)}{h}\, ds$$
$$+ \int_0^t \frac{f(\alpha + h, x(\alpha,s), s) - f(\alpha, x(\alpha,s), s)}{h}\, ds \tag{29}$$

We regress for the moment and consider the *linear* differential equation

$$\frac{dy}{dt} = f_2(\alpha, x(\alpha,t), t)\, y + f_1(\alpha, x(\alpha,t), t) \tag{30}$$

Here we use the notation $\partial f/\partial x = f_2$ and $\partial f/\partial \alpha = f_1$, where $f = f(\alpha,x,t)$ is the right-hand member in (28). Into these partial derivatives has been substituted the solution $x(\alpha,t)$ of (28). The hypotheses concerning $f(\alpha,x,t)$ imply that (30), in addition to being linear in y, is continuous in t for $0 \leq t \leq b$ and therefore possesses a (unique) solution satisfying $y = 0$ for $t = 0$. In fact, this particular solution satisfies the integral equation

$$y(t) = \int_0^t [f_2(\alpha, x(\alpha,s), s)\, y(s) + f_1(\alpha, x(\alpha,s), s)]\, ds \quad (31)$$

and we shall show that $y(t)$ is the limit as $h \to 0$ of the difference quotient in (29). To see this, we subtract (31) from (29) and express the result in the form

$$\frac{x(\alpha + h, t) - x(\alpha,t)}{h} - y(t)$$

$$= \int_0^t \left[\frac{f(\alpha + h, x(\alpha + h, s), s) - f(\alpha + h, x(\alpha,s), s)}{h} \right.$$

$$\left. - f_2(\alpha + h, x(\alpha,s), s)\, y(s) \right] ds$$

$$+ \int_0^t \left[\frac{f(\alpha + h, x(\alpha,s), s) - f(\alpha, x(\alpha,s), s)}{h} - f_1(\alpha, x(\alpha,s), s) \right] ds$$

$$+ \int_0^t [f_2(\alpha + h, x(\alpha,s), s) - f_2(\alpha, x(\alpha,s), s)]\, y(s)\, ds \quad (32)$$

Now using the mean-value theorem, we may write

$$f(\alpha + h, x(\alpha + h, s), s) - f(\alpha + h, x(\alpha,s), s)$$
$$= [f_2(\alpha + h, x(u,s), s) + \epsilon_1(s,h)][x(\alpha + h, s) - x(\alpha,s)] \quad (33)$$

and

$$f(\alpha + h, x(\alpha,s), s) - f(\alpha, x(\alpha,s), s) = [f_1(\alpha, x(\alpha,s), s) + \epsilon_2(s,h)]h$$
$$(34)$$

where $\epsilon_1(s,h)$ and $\epsilon_2(s,h)$ are continuous in s and tend to zero uniformly for $0 \leq s \leq b$, as $h \to 0$. Further, since f_2 is continuous, we have

$$f_2(\alpha + h, x(\alpha,s), s) - f_2(\alpha, x(\alpha,s), s) = \epsilon_3(s,h) \quad (35)$$

where $\epsilon_3(s,h)$ is also continuous in s and tends to zero uniformly for $0 \leq s \leq b$, as $h \to 0$. Introducing (33), (34), and (35) into

(32), the latter may be expressed in the form

$$\frac{x(\alpha + h,\, t) - x(\alpha,t)}{h} - y(t) = \int_0^t \left[\frac{x(\alpha + h,\, s) - x(\alpha,s)}{h} - y(s)\right]$$

$$[f_2(\alpha + h,\, x(\alpha,s),\, s) + \epsilon_1(s,h)]\, ds$$

$$+ \int_0^t \{\epsilon_2(s,h) + [\epsilon_1(s,h) + \epsilon_3(s,h)]\, y(s)\}\, ds \quad (36)$$

Now with

$$\Phi(t,h) = \left| \frac{x(\alpha + h,\, t) - x(\alpha,t)}{h} - y(t) \right|$$

$$\psi(s,h) = |f_2(\alpha + h,\, x(\alpha,s),\, s) + \epsilon_1(s,h)|$$

and
$$\rho(s,h) = |\epsilon_2(s,h) + [\epsilon_1(s,h) + \epsilon_3(s,h)]\, y(s)|$$

Eq. (36) leads to the inequality

$$\Phi(t,h) \le \int_0^t [\Phi(s,h)\, \psi(s,h) + \rho(s,h)]\, ds$$

Thus, by Lemma 1,

$$\Phi(t,h) \le c \exp \left\{ \int_0^t \left[\psi(s,h) + \frac{\rho(s,h)}{c}\right] ds \right\} \quad (37)$$

for every positive number c. In particular if we let

$$c = c(h) = \max_{0 \le s \le b} |\rho(s,h)|$$

then (37) results in

$$\Phi(t,h) \le c(h) \exp \left\{ \int_0^t [\psi(s,h) + 1]\, ds \right\} \quad (38)$$

Clearly $c(h) \to 0$ as $h \to 0$, and since at the same time, $\psi(s,h) \to |f_2(\alpha,\, x(\alpha,s),\, s)|$, (38) implies that $\Phi(t,h) \to 0$, in fact, uniformly for $0 \le t \le b$. Thus

$$\lim_{h \to 0} \frac{x(\alpha + h,\, t) - x(\alpha,t)}{h} = y(t)$$

uniformly for $0 \le t \le b$. This not only shows that $x(\alpha,t)$ is continuously differentiable with respect to α but that the derivative $dx\,(\alpha,t)/d\alpha$ is the unique solution $y(t)$ of the linear equation (30) which vanishes for $t = 0$. Equation (30) is known as the *equation of first variation* of (28).

To see that $dx\,(\alpha,t)/dt$ is continuously differentiable with respect to both α and t, one need merely observe that $f(\alpha,\,x(\alpha,t),\,t)$ is continuously differentiable with respect to both α and t. For by (28), the latter is already $dx\,(\alpha,t)/dt$. Finally, by induction, one extends the result to an arbitrary (finite) number of variables α and to higher-order derivatives.

EXERCISES

1. By introducing a suitable artifice, use Theorem 2 to show that the solution $x(t)$ of (28) satisfying the initial condition $x(t_0) = c$ is continuously differentiable with respect to t_0.

2. Prove Lemma 1.

3. Show that the derivative with respect to $\alpha_k(1 \le k \le j)$ of the solution $\bar{x}(t)$ of the initial-value problem (26) satisfies the linear vector differential equation $d\bar{y}/dt = \bar{f}_{\bar{x}}(\bar{x}(t),\,t)\,\bar{y} + \bar{f}_{\alpha_k}\,(\bar{x}(t),\,t)$, where the Jacobian matrix

$$\bar{f}_{\bar{x}} = \frac{d\bar{f}}{d\bar{x}} = \begin{pmatrix} \dfrac{\partial f_1}{\partial x_1} & \dfrac{\partial f_1}{\partial x_2} & \cdots & \dfrac{\partial f_1}{\partial x_n} \\ \dfrac{\partial f_2}{\partial x_1} & \dfrac{\partial f_2}{\partial x_2} & & \dfrac{\partial f_2}{\partial x_n} \\ \cdots & & & \cdots \\ \dfrac{\partial f_n}{\partial x_1} & \dfrac{\partial f_n}{\partial x_2} & \cdots & \dfrac{\partial f_n}{\partial x_n} \end{pmatrix}$$

and the partial derivative

$$\bar{f}_{\alpha_k} = \frac{\partial \bar{f}}{\partial \alpha_k} = \begin{pmatrix} \dfrac{\partial f_1}{\partial \alpha_k} \\ \dfrac{\partial f_2}{\partial \alpha_k} \\ \cdots \\ \dfrac{\partial f_n}{\partial \alpha_k} \end{pmatrix}$$

are each evaluated for $\bar{x} = \bar{x}(t)$. What is the appropriate initial condition for $\bar{y}$?

4. What linear vector differential equation is satisfied by the derivative with respect to t of the solution $\bar{x}(t)$ of (26)?   Illustrate for the scalar case (28).

5. What linear vector differential equation is satisfied by the derivative with respect to a component of $\bar{c}$ of the solution $\bar{y}(t)$ of (27)?   Illustrate for the scalar case (28).

6. What linear differential equation is satisfied by the derivative with respect to k of a solution u of the satellite equation $d^2u/dt^2 + u = ku^2$?   Obtain several terms of a power-series expansion of a periodic solution of the derived equation and compare with the k derivative of the expansion (19).

7. Using Exercise 5, show that the matrix

$$X = \frac{d\bar{x}}{d\bar{c}} = \begin{pmatrix} \dfrac{\partial x_1}{\partial c_1} & \dfrac{\partial x_1}{\partial c_2} & \cdots & \dfrac{\partial x_1}{\partial c_n} \\[2mm] \dfrac{\partial x_2}{\partial c_1} & \dfrac{\partial x_2}{\partial c_2} & \cdots & \dfrac{\partial x_2}{\partial c_n} \\[1mm] \cdot & \cdot & \cdots & \cdot \\[1mm] \dfrac{\partial x_n}{\partial c_1} & \dfrac{\partial x_n}{\partial c_2} & \cdots & \dfrac{\partial x_n}{\partial c_n} \end{pmatrix}$$

where $\bar{x} = \bar{x}(t)$ is the solution of (26) satisfying the initial condition $\bar{x}(0) = \bar{c}$, satisfies the linear matrix equation

$$\frac{dX}{dt} = \bar{f}_{\bar{x}}(\bar{x}(t),\, t)\ X$$

where $\bar{f}_{\bar{x}}$ is the Jacobian matrix of Exercise 3.   For t fixed, the determinant $|X|$ is the *Jacobian* of the transformation which maps each initial vector $\bar{c}$ onto the corresponding point $\bar{x}(t)$ along that trajectory which, in turn, has $\bar{c}$ as initial value for $t = 0$. Illustrate for the scalar case (28) and show that, as a function of t, the Jacobian J satisfies the linear differential equation $dJ/dt = f_x(\alpha,\, x(t),\, t)J$, where $f_x = \partial f/\partial x$ and $f = f(\alpha,x,t)$ is the right-hand member of (28).   Express J as a definite integral in t.

Chapter 4

PROPERTIES OF LINEAR SYSTEMS

1. Bases and the Principal Matrix Solution

Suppose that an n-by-n-matrix function $A(t)$ is continuous for $0 \leq t \leq b, b > 0$. Then for any n vector $\bar{c}$, there exists a unique solution of the linear initial-value problem

$$\frac{d\bar{x}}{dt} = A(t) \, \bar{x} \tag{1}$$

$$\bar{x}(0) = \bar{c} \tag{2}$$

for $0 \leq t \leq b$. This follows from the general existence-uniqueness theorems of Chap. 2. As one varies the initial vector $\bar{c}$ in n-space, one obtains all solutions (for $0 \leq t \leq b$) of the differential equation (1). In fact, the family of solutions of (1) form an n-dimensional vector space which is isomorphic to the n-dimensional vector space of initial vectors. Because there are bases which generate all $\bar{c}$ in n-space, there are also bases which generate all solutions of (1). We shall introduce a particularly important base which is to be used throughout.

If $\bar{y}^{(1)}, \bar{y}^{(2)}, \ldots, \bar{y}^{(n)}$ is any base of the solutions of (1), then each solution $\bar{x}$ of (1) is a linear (unique) combination of these basic vectors. That is, there exist suitable (unique) scalar constants $\alpha_1, \alpha_2, \ldots, \alpha_n$ such that

$$\bar{x}(t) = \alpha_1 \bar{y}^{(1)}(t) + \alpha_2 \bar{y}^{(2)}(t) + \cdots + \alpha_n \bar{y}^{(n)}(t) \tag{3}$$

holds identically in $t(0 \leq t \leq b)$. If $\bar{c}^{(1)}$, $\bar{c}^{(2)}$, . . . , $\bar{c}^{(n)}$ are the initial values (vectors) of the basic vectors, $\bar{y}^{(1)}$, $\bar{y}^{(2)}$, . . . , $\bar{y}^{(n)}$ respectively, then the initial-value (vector) of $\bar{x}$ is

$$\bar{c} = \alpha_1\bar{c}^{(1)} + \alpha_2\bar{c}^{(2)} + \cdots + \alpha_n\bar{c}^{(n)} \tag{4}$$

For the study of the initial-value problem (1), (2), we shall find it convenient to use the very special base determined by the initial vectors

$$\bar{c}^{(1)} = \begin{pmatrix} 1 \\ 0 \\ 0 \\ \cdots \\ 0 \end{pmatrix}, \ \bar{c}^{(2)} = \begin{pmatrix} 0 \\ 1 \\ 0 \\ \cdots \\ 0 \end{pmatrix}, \ \ldots, \ \bar{c}^{(n)} = \begin{pmatrix} 0 \\ 0 \\ \cdots \\ 0 \\ 1 \end{pmatrix} \tag{5}$$

The corresponding basic solutions $\bar{y}^{(1)}$, $\bar{y}^{(2)}$, . . . , $\bar{y}^{(n)}$ are not always the most natural nor the simplest to obtain, but they are particularly convenient for the study of the initial-value problem. For example, by (5), it is clear that the scalars α_1, α_2, . . . , α_n then become the components of $\bar{c}$, i.e.; the right-hand side of (4) becomes the vector

$$\begin{pmatrix} \alpha_1 \\ \alpha_2 \\ \cdots \\ \alpha_n \end{pmatrix}$$

Thus, the general solution of (1) is

$$\bar{x} = c_1\bar{y}^{(1)} + c_2\bar{y}^{(2)} + \cdots + c_n\bar{y}^{(n)} \tag{6}$$

where
$$\bar{c} = \begin{pmatrix} c_1 \\ c_2 \\ \cdots \\ c_n \end{pmatrix} \tag{7}$$

is the initial value (vector) of $\bar{x}$. It is convenient to introduce a single symbol for the base $\bar{y}^{(1)}$, $\bar{y}^{(2)}$, . . . , $\bar{y}^{(n)}$ and this we do in a natural way with the n-by-n matrix

$$Y = (\bar{y}^{(1)} \quad \bar{y}^{(2)} \quad \cdots \quad \bar{y}^{(n)}) = \begin{pmatrix} y_1^{(1)} & y_1^{(2)} & \cdots & y_1^{(n)} \\ y_2^{(1)} & y_2^{(2)} & \cdots & y_2^{(n)} \\ \cdots & \cdots & \cdots & \cdots \\ y_n^{(1)} & y_n^{(2)} & \cdots & y_n^{(n)} \end{pmatrix} \tag{8}$$

which consists of a row of the column vectors $\bar{y}^{(1)}, \bar{y}^{(2)}, \ldots, \bar{y}^{(n)}$. Here, and throughout the remaining chapters, $\bar{y}^{(1)}, \bar{y}^{(2)}, \ldots, \bar{y}^{(n)}$ will always denote the n particular solutions of (1) which assume the initial values (5) for $t = 0$. Similarly, Y will always denote the matrix array (8) of these particular solutions. Equations (6) and (7) are compactly reformulated in the single equation

$$\bar{x} = Y\bar{c} \tag{9}$$

Then the general solution of (1) is expressed as the product of the fixed n-by-n-matrix function Y and an arbitrary (constant) column vector $\bar{c}$. In this way, the initial-value vector $\bar{c}$ of a solution $\bar{x}$ is exhibited in a direct and compact fashion. For any such pair, we have, from (9),

$$\frac{d\bar{x}}{dt} = \frac{dY}{dt}\bar{c}$$

More generally, if we consider n arbitrary solutions (not necessarily distinct nor linearly independent) $\bar{x}^{(1)}, \bar{x}^{(2)}, \ldots, \bar{x}^{(n)}$ of (1) corresponding to n arbitrary initial vectors $\bar{c}^{(1)}, \bar{c}^{(2)}, \ldots, \bar{c}^{(n)}$, then we have, from (9), the n vector equations

$$\frac{d\bar{x}^{(j)}}{dt} = \frac{dY}{dt}\bar{c}^{(j)} \qquad j = 1, 2, \ldots, n \tag{10}$$

The n vector equations (n^2 scalar equations) represented by (10) are equivalent to the single n-by-n-matrix equation

$$\frac{dX}{dt} = \frac{dY}{dt}C \tag{11}$$

where
$$X = (\bar{x}^{(1)} \quad \bar{x}^{(2)} \quad \cdots \quad \bar{x}^{(n)}) \tag{12}$$
and
$$C = (\bar{c}^{(1)} \quad \bar{c}^{(2)} \quad \cdots \quad \bar{c}^{(n)}) \tag{13}$$

Thus the matrix analogue of (9) becomes

$$X = YC \tag{14}$$

where C is called the *initial condition matrix*. Equation (14) is merely a compact expression for n vector solutions of (1) in terms of the corresponding n initial vectors. On the other hand, since for each $j = 1, 2, \ldots, n$, $\bar{x}^{(j)}$ is a solution of (1), we have

$$\frac{d\bar{x}^{(j)}}{dt} = A(t) \, \bar{x}^{(j)} \tag{15}$$

and the n differential equations (15) for $j = 1, 2, \ldots, n$ may be written compactly as the single matrix differential equation

$$\frac{dX}{dt} = A(t) \, X \tag{16}$$

Equation (16) is called the *associated matrix equation*. Every matrix solution of (16) corresponds to n vector solutions of (1). Conversely, every set of n vector solutions of (1) corresponds to one matrix solution of (16). Equation (16) is equivalent to n^2 simultaneous scalar differential equations in which there is considerable algebraic redundancy. In fact, if the equivalent n^2 scalar equations are written in the usual form of a column vector, [rather than the square array implied by (16)], then the n^2-by-n^2-system matrix may be expressed in the form

$$E(t) = \begin{pmatrix} A(t) & (0) & (0) & \cdots & (0) \\ (0) & A(t) & (0) & \cdots & (0) \\ \cdots & \cdots & \cdots & \cdots & \cdots \\ (0) & (0) & \cdots & (0) & A(t) \end{pmatrix}$$

with the n-by-n matrix $A(t)$ appearing along the principal diagonal as a submatrix. The corresponding vector differential equation

$$\frac{d\bar{z}}{dt} = E(t) \, \bar{z} \tag{17}$$

is a linear vector system whose coefficient matrix $E(t)$ is continuous. The general existence-uniqueness theorems apply to (17) and thus also to the equivalent matrix system (16).

The general solution of (16) is given by (14), where Y is the particular solution of (16) satisfying the initial condition

$$Y(0) = \begin{pmatrix} 1 & 0 & \cdots & 0 \\ 0 & 1 & \cdots & 0 \\ 0 & 0 & \cdots & 0 \\ \vdots & & & \vdots \\ 0 & \cdots & \cdots & 0 \\ 0 & 0 & \cdots & 1 \end{pmatrix} = I \text{ (the identity matrix)}$$

$Y(t)$ is called the *principal matrix solution*.

EXAMPLE 1

Consider the equation $d^2x/dt^2 + x = 0$, or rather, the equivalent system

$$\frac{dx_1}{dt} = x_2$$
$$\frac{dx_2}{dt} = -x_1$$

(18)

The vector version of (18) is

$$\frac{d\bar{x}}{dt} = A\bar{x}$$

(19)

where

$$A = \begin{pmatrix} 0 & 1 \\ -1 & 0 \end{pmatrix}$$

We have the elementary solutions

$$\bar{y}^{(1)} = \begin{pmatrix} \cos t \\ -\sin t \end{pmatrix}$$

and

$$\bar{y}^{(2)} = \begin{pmatrix} \sin t \\ \cos t \end{pmatrix}$$

If we let

$$Y = (\bar{y}^{(1)} \quad \bar{y}^{(2)}) = \begin{pmatrix} \cos t & \sin t \\ -\sin t & \cos t \end{pmatrix}$$

then Y is the principal matrix solution. Indeed, $Y(0)$ is the identity matrix. The general solution of the matrix equation $dX/dt = AX$ is

$$X = YC = \begin{pmatrix} \cos t & \sin t \\ -\sin t & \cos t \end{pmatrix} \begin{pmatrix} c_{11} & c_{12} \\ c_{21} & c_{22} \end{pmatrix}$$

for arbitrary c_{11}, c_{12}, c_{21}, and c_{22} and the general solution of (19) is

$$\bar{x} = Y\bar{c} = \begin{pmatrix} \cos t & \sin t \\ -\sin t & \cos t \end{pmatrix} \begin{pmatrix} c_1 \\ c_2 \end{pmatrix} = c_1 \begin{pmatrix} \cos t \\ -\sin t \end{pmatrix} + c_2 \begin{pmatrix} \sin t \\ \cos t \end{pmatrix}$$
$$= c_1 \bar{y}^{(1)} + c_2 \bar{y}^{(2)}$$

for arbitrary c_1 and c_2.

EXAMPLE 2

Consider the equation

$$\frac{d^2x}{dt^2} - \lambda^2 x = 0 \tag{20}$$

and its vector equivalent

$$\frac{d\bar{x}}{dt} = A\bar{x} \tag{21}$$

where

$$A = \begin{pmatrix} 0 & 1 \\ \lambda^2 & 0 \end{pmatrix}$$

As is well known, $x^{(1)} = e^{\lambda t}$ and $x^{(2)} = e^{-\lambda t}$ are solutions of (20) and form a base for all solutions. Hence

$$\bar{x}^{(1)} = \begin{pmatrix} e^{\lambda t} \\ \lambda e^{\lambda t} \end{pmatrix} \quad \text{and} \quad \bar{x}^{(2)} = \begin{pmatrix} e^{-\lambda t} \\ -\lambda e^{-\lambda t} \end{pmatrix}$$

form a base for all solutions of (21). On the other hand,

$$\bar{x}^{(1)}(0) = \begin{pmatrix} 1 \\ \lambda \end{pmatrix} \quad \text{and} \quad \bar{x}^{(2)}(0) = \begin{pmatrix} 1 \\ -\lambda \end{pmatrix} \tag{22}$$

so that this base does *not* correspond to the principal matrix solution. Considering the initial vectors (22), it is clear that

$\bar{y}^{(1)} = [\bar{x}^{(1)} + \bar{x}^{(2)}]/2$ and $\bar{y}^{(2)} = [\bar{x}^{(1)} - \bar{x}^{(2)}]/2\lambda$ are the desired basic vectors. Interestingly enough, these correspond to the familiar hyperbolic solutions $\cosh \lambda t$ and $(1/\lambda) \sinh \lambda t$ of (20). The principal matrix solution is

$$Y = \begin{pmatrix} \cosh \lambda t & (1/\lambda) \sinh \lambda t \\ \sinh \lambda t & \cosh \lambda t \end{pmatrix}$$

and the general solution of (21) is

$$\bar{x} = \begin{pmatrix} \cosh \lambda t & (1/\lambda) \sinh \lambda t \\ \sinh \lambda t & \cosh \lambda t \end{pmatrix} \begin{pmatrix} c_1 \\ c_2 \end{pmatrix} = c_1 \bar{y}^{(1)} + c_2 \bar{y}^{(2)}$$

for arbitrary c_1 and c_2.

EXAMPLE 3

Consider the nth-order linear equation

$$\frac{d^n x}{dt^n} + b_n \frac{d^{n-1} x}{dt^{n-1}} + \cdots + b_1 x = 0 \tag{23}$$

where $b_1, b_2, \ldots, b_n$ are continuous functions of t. Let $x^{(1)}, x^{(2)}, \ldots, x^{(n)}$ be an arbitrary family of n solutions of (23). Corresponding to these n scalar solutions there are n vector solutions $\bar{x}^{(1)}, \bar{x}^{(2)}, \ldots, \bar{x}^{(n)}$ of

$$\frac{d\bar{x}}{dt} = A\bar{x} \tag{24}$$

where

$$\bar{x} = \begin{pmatrix} x \\ \dfrac{dx}{dt} \\ \cdots \\ \dfrac{d^{n-1} x}{dt^{n-1}} \end{pmatrix}$$

and

$$A = \begin{pmatrix} 0 & 1 & 0 & 0 & \cdots & 0 \\ 0 & 0 & 1 & 0 & \cdots & 0 \\ \cdot & \cdot & \cdot & \cdot & \cdots & \cdot \\ 0 & 0 & \cdots & \cdots & 0 & 1 \\ -b_1 & -b_2 & \cdots & \cdots & \cdots & -b_n \end{pmatrix}$$

The first component of $\bar{x}^{(j)}$ is $x^{(j)}$, $j = 1, 2, \ldots, n$. Let $X = (\bar{x}^{(1)}\bar{x}^{(2)} \cdots \bar{x}^{(n)})$ be the corresponding matrix solution of

$$\frac{dX}{dt} = AX \tag{25}$$

and $w = |X| =$ determinant of X. Evidently

$$w = \begin{vmatrix} x^{(1)} & x^{(2)} & \cdots & x^{(n)} \\ \dfrac{dx^{(1)}}{dt} & \dfrac{dx^{(2)}}{dt} & \cdots & \dfrac{dx^{(n)}}{dt} \\ \cdots & \cdots & \cdots & \cdots \\ \dfrac{d^{(n-1)}x^{(1)}}{dt^{n-1}} & \dfrac{d^{(n-1)}x^{(2)}}{dt^{n-1}} & \cdots & \dfrac{d^{(n-1)}x^{(n)}}{dt^{n-1}} \end{vmatrix}$$

This determinant is commonly referred to as the *Wronskian* of the n solutions $x^{(1)}, x^{(2)}, \ldots, x^{(n)}$. In particular, it is well known that the n solutions are linearly independent if and only if the Wronskian is different from zero. Thus they are independent if the corresponding matrix solution $X(t)$ is nonsingular. Our next theorem will establish the converse.

♦ ♦ ♦ ♦ ♦ ♦ ♦

Consider the matrix equation

$$\frac{dX}{dt} = A(t) X \tag{26}$$

where $A = (a_{ij})$, and let $w = |X|$ be the determinant of a matrix solution of (26). We wish to discuss the nature of the scalar function $w = w(t) = |X(t)|$. The determinant $|X|$ consists of sums of products with each product containing n factors (consisting of elements of the matrix X, exactly one factor from each row and each column) multiplied by $+1$ or -1. Thus, from the product rule for differentiation, it follows that the derivative dw/dt consists of the sum of n similar determinants, within each of which exactly one row of elements has been replaced by corresponding derivatives. Thus

$$\frac{dw}{dt} = \begin{vmatrix} \dfrac{dx_{11}}{dt} & \dfrac{dx_{12}}{dt} & \cdots & \dfrac{dx_{1n}}{dt} \\ x_{21} & x_{22} & \cdots & x_{2n} \\ \cdots & \cdots & \cdots & \cdots \\ x_{n1} & x_{n2} & \cdots & x_{nn} \end{vmatrix} + \begin{vmatrix} x_{11} & x_{12} & \cdots & x_{1n} \\ \dfrac{dx_{21}}{dt} & \dfrac{dx_{22}}{dt} & & \dfrac{dx_{2n}}{dt} \\ x_{31} & x_{32} & \cdots & x_{3n} \\ \cdots & \cdots & \cdots & \cdots \\ x_{n1} & x_{n2} & \cdots & x_{nn} \end{vmatrix}$$

$$+ \cdots + \begin{vmatrix} x_{11} & x_{12} & \cdots & x_{1n} \\ \cdots & \cdots & \cdots & \cdots \\ \cdots & \cdots & \cdots & \cdots \\ x_{(n-1)1} & x_{(n-1)2} & \cdots & x_{(n-1)n} \\ \dfrac{dx_{n1}}{dt} & \dfrac{dx_{n2}}{dl} & \cdots & \dfrac{dx_{nn}}{dt} \end{vmatrix} \tag{27}$$

Since the columns of X are solutions of the vector equation

$$\frac{d\bar{x}}{dt} = A(t)\,\bar{x} \tag{28}$$

we have, for the derivatives in the first determinant of (27)

$$\frac{dx_{11}}{dt} = a_{11}x_{11} + a_{12}x_{21} + \cdots + a_{1n}x_{n1}$$

$$\frac{dx_{12}}{dt} = a_{11}x_{12} + a_{12}x_{22} + \cdots + a_{1n}x_{n2} \tag{29}$$

$$\cdots \cdots \cdots \cdots \cdots \cdots$$

$$\frac{dx_{1n}}{dt} = a_{11}x_{1n} + a_{12}x_{2n} + \cdots + a_{1n}x_{nn}$$

If the derivatives appearing in the first determinant in (27) are replaced by their corresponding expressions in (29), we obtain

$$\begin{vmatrix} \displaystyle\sum_{i=1}^{n} a_{1i}x_{i1} & \displaystyle\sum_{i=1}^{n} a_{1i}x_{i2} & \cdots & \displaystyle\sum_{i=1}^{n} a_{1i}x_{in} \\ x_{21} & x_{22} & \cdots & x_{2n} \\ \cdots & \cdots & \cdots & \cdots \\ x_{n1} & x_{n2} & \cdots & x_{nn} \end{vmatrix} \tag{30}$$

Now if we multiply the second row of (30) by a_{12} and subtract from the first row, the value of the determinant is left unchanged.

Similarly, if we multiply the ith row of (30) by a_{1i} $(2 \leq i \leq n)$ and subtract from the first, the value of the determinant is left unchanged. It follows, therefore, that the determinant

$$
\begin{vmatrix}
a_{11}x_{11} & a_{11}x_{12} & \cdots & a_{11}x_{1n} \\
x_{21} & x_{22} & \cdots & x_{2n} \\
\cdots & \cdots & \cdots & \cdots \\
x_{n1} & x_{n2} & \cdots & x_{nn}
\end{vmatrix}
\tag{31}
$$

has the same value as (30). But (31) has the same value as $a_{11}|X|$. Similar considerations of the kth determinant in (27) show that the value of the kth determinant is given by $a_{kk}|X|$. Thus (27) may be expressed in the form

$$
\frac{dw}{dt} = a_{11}w + a_{22}w + \cdots + a_{nn}w = \left(\sum_{k=1}^{n} a_{kk} \right) w \tag{32}
$$

where the scalar quantity $\sum_{k=1}^{n} a_{kk}$ is the *trace of the matrix* A. We have, therefore, the important Theorem 1.

THEOREM 1

If X is a solution of the matrix equation

$$
\frac{dX}{dt} = A(t) \, X \tag{33}
$$

for $0 \leq t \leq b$, $b > 0$, then

$$
|X(t)| = |X(0)| \exp \left[\int_0^t \text{trace } A(s) \, ds \right] \tag{34}
$$

for $0 \leq t \leq b$.

PROOF: The general solution of the scalar equation (32) has the form $w = w(0) \exp \left[\int_0^t \text{trace } A(s) \, ds \right]$ which is (34).

The importance of (34) lies in the fact that if $|X(0)| \neq 0$, i.e., the initial condition matrix $C = X(0)$ is nonsingular, then the matrix solution $X(t)$ of (33) remains nonsingular. In particular, the principal matrix solution $Y(t)$ is always nonsingular since

$Y(0) = I$. In vector language, Theorem 1 states that a set of n solutions of (1) remain linearly independent if they are initially linearly independent and that they remain linearly dependent if they are initially linearly dependent.

For the nth-order equation in Example 3, the trace of A is merely the negative of the coefficient b_n. Thus, the Wronskian satisfies $w(t) = |X(t)| = |X(0)| \exp \left[\int_0^t - b_n(s) \, ds \right]$, a well-known result for nth-order linear differential equations.

EXERCISES

1. Show that the family of solutions of (1) form a vector space. Show that this vector space of solutions is isomorphic to the vector space of initial values.

2. Explain, in detail, the equivalence of the vector system (17) and the matrix system (16). What solution of (17) corresponds to the principal matrix solution of (16)?

3. State and prove (without reference to vector systems) an existence-uniqueness theorem for the matrix system (16).

4. Obtain the principal matrix solution for the associated matrix equation of the general linear second-order equation $d^2x/dt^2 + k \, dx/dt + \omega^2 x = 0$ with constant coefficients. Consider the various cases discussed in Chap. 1.

5. Let each of x_1 and x_2 be a solution of the above linear second-order equation. Multiply $d^2x_1/dt^2 + k \, dx_1/dt + \omega^2 x_1 = 0$ by x_2 and $d^2x_2/dt^2 + k \, dx_2/dt + \omega^2 x_2 = 0$ by x_1; subtract and show that the quantity $P = x_2^2 d(x_1/x_2)/dt$ satisfies the first-order equation $dP/dt + kP = 0$. Thus show that if $x_2(0) \neq 0$, then $x_1 = c_1 x_2 + c_2 x_2 \int_0^t (e^{-ks}/x_2^2) \, ds$ for suitable constants c_1 and c_2. Discuss the independence of the solutions x_1 and x_2.

6. Using the notation of Exercise 5, show that the quantity $Q = x_1 \, dx_2/dt - x_2 \, dx_1/dt$ satisfies the first order linear equation $dQ/dt + kQ = 0$. Show that this leads to (34) for this special case.

7. Generalize the technique in Exercise 6, and derive (34) for the general nth-order linear equation (23).

8. Consider the second-order equation

$$\frac{d^2x}{dt^2} + k\frac{dx}{dt} + \omega^2 x = 0 \tag{i}$$

where each of k and ω^2 is a constant. Illustrate Theorem 1 for this special case. For t fixed, define a transformation of the phase-plane vectors onto themselves such that

$$\begin{pmatrix} c_1 \\ c_2 \end{pmatrix} \text{ maps to } \begin{pmatrix} x \\ \dfrac{dx}{dt} \end{pmatrix} = \begin{pmatrix} x_1 \\ x_2 \end{pmatrix}$$

where $x = c_1$ and $dx/dt = c_2$ for $t = 0$. The Jacobian of the transformation is given by

$$J = \begin{vmatrix} \dfrac{\partial x_1}{\partial c_1} & \dfrac{\partial x_2}{\partial c_1} \\ \dfrac{\partial x_1}{\partial c_2} & \dfrac{\partial x_2}{\partial c_2} \end{vmatrix}$$

and depicts the changes in area under the transformation. Using (i), obtain the relations

$$\frac{d}{dt}\left(\frac{\partial x_1}{\partial c_1}\right) = \frac{\partial x_2}{\partial c_1} \qquad \frac{d}{dt}\left(\frac{\partial x_1}{\partial c_2}\right) = \frac{\partial x_2}{\partial c_2}$$

$$\frac{d}{dt}\left(\frac{\partial x_2}{\partial c_1}\right) = -\omega^2\frac{\partial x_1}{\partial c_1} - k\frac{\partial x_2}{\partial c_1} \qquad \frac{d}{dt}\left(\frac{\partial x_2}{\partial c_2}\right) = -\omega^2\frac{\partial x_1}{\partial c_2} - k\frac{\partial x_2}{\partial c_2}$$

and then show that J, as a function of t, satisfies the differential equation $dJ/dt = -kJ$. Explain why this proves Theorem 1 for the special case (i).

9. Using the technique outlined in Exercise 8, show that the corresponding Jacobian J for the nonlinear equation

$$\frac{d^2x}{dt^2} + f(x)\frac{dx}{dt} + \omega^2 x = 0$$

is given by $J = \exp\left[-\int_0^t f(x(s))\, ds\right]$.

2. The Linear Inhomogeneous Equation (The Adjoint Equation)

Consider the initial-value problem

$$\frac{d\bar{x}}{dt} = A(t)\,\bar{x} + \bar{f}(t) \tag{35}$$

$$\bar{x}(0) = \bar{c} \tag{36}$$

where $A(t)$ and $\bar{f}(t)$ are continuous for $0 \leq t \leq b$, $b > 0$. There exists a unique solution of (35), (36) for $0 \leq t \leq b$. In fact, we shall obtain an explicit expression for this unique solution.

Let $Z(t)$ be the matrix solution, satisfying $Z(0) = I$, of the equation

$$\frac{dZ}{dt} = -Z\,A(t) \tag{37}$$

where $A(t)$ is as in (35) and I is the identity matrix. The reduced associated matrix equation, corresponding to (35), is

$$\frac{dX}{dt} = A(t)\,X \tag{38}$$

and Eq. (37) is called the *adjoint* of (38). There is a simple algebraic relationship between any solution of (37) and any solution of (38). For, if we multiply (37) on the right by X and (38) on the left by Z and add, then the sum may be expressed in the form

$$\frac{d}{dt}(ZX) = 0$$

Hence, the product ZX is always a constant, i.e.,

$$ZX = C(\text{constant matrix}) \tag{39}$$

In the present case, Z is nonsingular and we may solve (30) to obtain X,

$$X = Z^{-1}C \tag{40}$$

For C arbitrary, this is the general solution of (38). On the other hand, $Z^{-1} = I$ for $t = 0$ and so, necessarily, Z^{-1} is the principal matrix solution $Y(t)$ of (38). Then, of course, $Y^{-1}(t)$ is the principal matrix solution of (37). We have, therefore, in addition to

$$\frac{dY}{dt} = A(t)\ Y$$

the reciprocal (adjoint) equation

$$\frac{dY^{-1}}{dt} = -Y^{-1}\ A(t) \tag{41}$$

for any principal matrix solution.

If Eq. (35) is multiplied on the left by $Y^{-1}(t)$ and Eq. (41) is multiplied on the right by $\bar{x}$, and if the two results are added, then the sum may be expressed in the form

$$\frac{d(Y^{-1}\bar{x})}{dt} = Y^{-1}(t)\ \bar{f}(t)$$

This in turn implies that

$$Y^{-1}(t)\ \bar{x}(t) = \bar{c} + \int_0^t Y^{-1}(s)\ \bar{f}(s)\ ds \tag{42}$$

The constant of integration in (42) is the initial vector $\bar{c}$, since $Y^{-1}(0) = I$. Thus

$$\bar{x}(t) = Y(t)\ \bar{c} + Y(t) \int_0^t Y^{-1}(s)\ \bar{f}(s)\ ds \tag{43}$$

is the solution of the initial-value problem (35), (36). The first term on the right is the general solution of the corresponding reduced equation, i.e., the corresponding homogeneous equation, and the second represents the response to the forcing function $\bar{f}(t)$ of the system initially at rest.

EXAMPLE 4

Using Example 1, we obtain the general solution of the equation $d^2x/dt^2 + x = f(t)$ in the form

$$\bar{x} = \begin{pmatrix} \cos t & \sin t \\ -\sin t & \cos t \end{pmatrix} \begin{pmatrix} c_1 \\ c_2 \end{pmatrix}$$
$$+ \begin{pmatrix} \cos t & \sin t \\ -\sin t & \cos t \end{pmatrix} \int_0^t \begin{pmatrix} \cos s & \sin s \\ -\sin s & \cos s \end{pmatrix}^{-1} \begin{pmatrix} 0 \\ f(s) \end{pmatrix} ds$$

where

$$\bar{x} = \begin{pmatrix} x \\ \dfrac{dx}{dt} \end{pmatrix}$$

It may be verified directly that

$$\begin{pmatrix} \cos s & \sin s \\ -\sin s & \cos s \end{pmatrix}^{-1} = \begin{pmatrix} \cos s & -\sin s \\ \sin s & \cos s \end{pmatrix}$$

Thus

$$\begin{pmatrix} \cos s & \sin s \\ -\sin s & \cos s \end{pmatrix}^{-1} \begin{pmatrix} 0 \\ f(s) \end{pmatrix} = \begin{pmatrix} -\sin s & f(s) \\ \cos s & f(s) \end{pmatrix}$$

and the expression for x becomes

$$x = c_1 \cos t + c_2 \sin t - \cos t \int_0^t \sin s \, f(s) \, ds$$
$$+ \sin t \int_0^t \cos s \, f(s) \, ds$$
$$= c_1 \cos t + c_2 \sin t + \int_0^t (\sin t \cos s - \sin s \cos t) \, f(s) \, ds$$
$$= c_1 \cos t + c_2 \sin t + \int_0^t \sin (t - s) \, f(s) \, ds$$

EXAMPLE 5

Using Example 2, we obtain the general solution of the equaton $d^2x/dt^2 - \lambda^2 x = f(t)$ in the form

$$\bar{x} = \begin{pmatrix} \cosh \lambda t & \dfrac{1}{\lambda} \sinh \lambda t \\ \lambda \sinh \lambda t & \cosh \lambda t \end{pmatrix} \begin{pmatrix} c_1 \\ c_2 \end{pmatrix}$$
$$+ \begin{pmatrix} \cosh \lambda t & \dfrac{1}{\lambda} \sinh \lambda t \\ \lambda \sinh \lambda t & \cosh \lambda t \end{pmatrix} \begin{pmatrix} \int_0^t \dfrac{-\sinh \lambda s}{\lambda} \, f(s) \, ds \\ \int_0^t \cosh \lambda s \quad f(s) \, ds \end{pmatrix}$$

where we have used

$$\begin{pmatrix} \cosh \lambda t & \dfrac{1}{\lambda}\sinh \lambda t \\ \lambda \sinh \lambda t & \cosh \lambda t \end{pmatrix}^{-1} = \begin{pmatrix} \cosh \lambda t & -\dfrac{1}{\lambda}\sinh \lambda t \\ -\lambda \sinh \lambda t & \cosh \lambda t \end{pmatrix}$$

Thus x is given by

$$x(t) = c_1 \cosh \lambda t + c_2 \frac{\sinh \lambda t}{\lambda} - \cosh \lambda t \int_0^t \frac{\sinh \lambda s}{\lambda} f(s)\, ds$$

$$+ \frac{\sinh \lambda t}{\lambda} \int_0^t \cosh \lambda s\, f(s)\, ds$$

$$= c_1 \cosh \lambda t + c_2 \frac{\sinh \lambda t}{\lambda}$$

$$+ \frac{1}{\lambda}\int_0^t (\sinh \lambda t \cosh \lambda s - \sinh \lambda s \cosh \lambda t)\, f(s)\, ds$$

$$= c_1 \cosh \lambda t + c_2 \frac{\sinh \lambda t}{\lambda} + \frac{1}{\lambda}\int_0^t \sinh \lambda(t - s)\, f(s)\, ds$$

• • • • • • • •

If A is a *constant* matrix, then the principal matrix solution $Y(t)$ of (38) satisfies

a. $Y(t + s) = Y(t)\, Y(s)$, for all s and t with $0 \le s \le b$ and $0 \le t \le b$.

In fact, for s fixed, $X(t) = Y(t + s)$ is a solution of (38) satisfying $X(0) = Y(s) = C$. Hence, from (14), $Y(t + s)$ may be expressed in the form $Y(t + s) = Y(t)\, C = Y(t)\, Y(s)$, which is property a above. For constant A, therefore, $Y(t)$ shares a familiar property with the scalar exponential $e^{\alpha t}$ [namely, $e^{\alpha(t+s)} = e^{\alpha t}e^{\alpha s}$] which is the solution of $dx/dt = \alpha x$ satisfying $x(0) = 1$. We denote by e^{At}, the principal matrix solution $Y(t)$ whenever A is a *constant* matrix, and the analogy with the scalar case is then complete. It is possible to define the exponential of a matrix as the sum of an infinite series of matrices, so that the familiar power-series expansion of $e^{\alpha t}$ about $t = 0$ carries over formally to e^{At}. It should be observed that property a implies that the matrices $Y(t)$ and $Y(s)$ commute.

Another property of interest for constant A is the following. Clearly, the matrix $X^{(1)}(t) = Y(t)\,A$ is a solution of (38) satisfying $X^{(1)}(0) = A$. On the other hand, if we multiply (38) on the left by A, we have

$$\frac{dAX}{dt} = A(AX)$$

and hence, in particular, the matrix $X^{(2)}(t) = A\,Y(t)$ appears as a solution of (38) satisfying the initial condition $X^{(2)}(0) = A$. Thus, since the solution of this initial-value problem is unique, we have

b. $A\,Y(t) = Y(t)\,A$

This is to say, $Y(t)$ commutes with A. Property b would be obvious from a power-series representation of $Y(t) = e^{At}$. Clearly, the matrix $X^{(3)}(t) = Y(-t)$ satisfies the equation $dX^{(3)}/dt = -AX^{(3)}$ and by property b, therefore, also the adjoint equation $dX^{(3)}/dt = -X^{(3)}A$. Since $X^{(3)}(0) = Y(0) = I$,

$$X^{(3)}(t) = Y(-t)$$

is the principal matrix solution of the adjoint equation and, in fact, the inverse of $Y(t)$ [see (41)]. Thus, we have

c. $Y^{-1}(t) = Y(-t)$

This also follows from property a since for $s = -t$, a yields $Y(t)\,Y(-t) = Y(0) = I$.

Because of properties a and c, the general solution (43) may be written in the form

$$\bar{x}(t) = Y(t)\,\bar{c} + \int_0^t Y(t-s)\,\bar{f}(s)\,ds \tag{44}$$

This is the general solution of an inhomogeneous linear system with constant coefficients. It involves a convolution integral which convolutes the principal matrix solution of the corresponding homogeneous system with the forcing function of the inhomogeneous system. The final expressions for the solutions in Examples 4 and 5 were each of the form (44).

EXERCISES

1. Show that each *row* vector $\bar{z}'$ of the matrix Z in (37) satisfies the vector equation

$$d\bar{z}'/dt = -\bar{z}'A \tag{i}$$

where $\bar{z}'A$ is the usual row by column product of a row vector and a square matrix.   Equation (i) is called the *adjoint* of the vector equation (1).

2. Determine the adjoint equations of each of (19), (21), and (24).   Give the scalar version of each of these adjoint systems.

3. Let $A'(t)$ denote the transpose of the matrix $A(t)$.   Show that the transpose of the matrix solution Z of (37) satisfies the equation $dZ'/dt = -A'(t)\,Z'$.   Hence, using Exercise 1, show that if a row vector $\bar{z}'$ of the matrix Z is written as a column vector $\bar{z}$, then $d\bar{z}/dt = -A'(t)\,\bar{z}$.   If the matrix $A(t)$ is skew symmetric, i.e., $A(t) = -A'(t)$, then the system (38) is said to be *self-adjoint*.   What special geometric characteristics of solutions of (1) are expressed by (39) for self-adjoint systems?   What special properties are possessed by the principal matrix solution $Y(t)$ of a self-adjoint system?

4. Verify that (43) is a solution of (35).

5. Derive the solutions in Examples 4 and 5 using the method of variation of parameters.

6. Using property a and Examples 1 and 2, derive some familiar trigonometric and hyperbolic identities.

7. Let A be a constant matrix.   Show that the matrix series $\sum_{k=0}^{\infty} A^k t^k / k!$ converges absolutely and uniformly on each interval $0 \leq t \leq b$, $b > 0$.   In fact, show that the series satisfies the differential equation $dX/dt = AX$ and then explain why it is necessarily the principal matrix solution.   Show directly that properties a, b, and c hold for the above series.

8. Let each of A and B be a constant n-by-n matrix.   Give two proofs that $e^{(A+B)t} = e^{At}e^{Bt}$ for all t, if and only if $AB = BA$.

9. Consider the absolutely and uniformly (for $0 \leq t \leq b$, $b > 0$) convergent series

$$\exp\left[\int_0^t A(s)\, ds\right] = \sum_{k=0}^{\infty} \left[\int_0^t A(s)\, ds\right]^k \frac{t^k}{k!}$$

Under what circumstances is this the principal matrix solution of (38)?

10. Interpret the convolution integral in (44) as the superposition of solutions of the homogeneous system

$$\frac{d\bar{x}}{dt} = A(t)\, \bar{x} \tag{ii}$$

For each fixed $s > 0$, determine the solution $\bar{x}$ of (ii) which satisfies the initial condition $\bar{x} = \bar{f}(s)$ for $t = s$ and then explain how to construct (44) by superposition.

11. Show that (44) may be written in the alternate form

$$\bar{x}(t) = Y(t)\, \bar{c} + \int_0^t Y(s)\, \bar{f}(t - s)\, ds$$

Interpret this form of the convolution integral as the superposition of solutions of (ii).

12. Which of the properties a, b, and c does the principal matrix solution share with *all* nonsingular matrix solutions of (38)? A nonsingular matrix solution of (35) is referred to as a *fundamental matrix solution*.

13. Let $X(t)$ be a fundamental matrix solution of (38). Show that

$$\bar{x}(t) = X(t)\, \bar{c} + X(t) \int_0^t X^{-1}(s)\, \bar{f}(s)\, ds$$

is a solution of (35) for any constant vector $\bar{c}$. Explain why this is a general solution of (38).

3. *Linear Homogeneous Equations with Constant Coefficients*

We now study, in detail, the special equation

$$\frac{d\bar{x}}{dt} = A\bar{x} \tag{45}$$

where A is a constant (real) matrix.   Much of what we shall do here reflects, for n-space, concepts which were introduced in Chap. 1 for two-space.   In fact, we begin by considering a linear transformation

$$\bar{u} = B\bar{x} \tag{46}$$

where B is a nonsingular (constant, not necessarily real) matrix. Equation (46) may be inverted to yield

$$\bar{x} = B^{-1}\bar{u} \tag{47}$$

We interpret (46) merely as a change of coordinates in a complex n-space and seek the corresponding differential equation in $\bar{u}$. For this purpose, the inverse mapping (47) is easier to treat than the direct mapping (46).   For from (47), we have

$$\frac{d\bar{x}}{dt} = B^{-1}\frac{d\bar{u}}{dt}$$

and from (45) $d\bar{x}/dt = AB^{-1}\bar{u}$.   Hence

$$\frac{d\bar{u}}{dt} = (BAB^{-1})\bar{u} = E\bar{u} \tag{48}$$

where $E = BAB^{-1}$ is an equivalent coefficient matrix in the $\bar{u}$ coordinates.   In matrix algebra, two matrices A and E are said to be *similar* if there exists a nonsingular matrix B such that

$$E = BAB^{-1} \tag{49}$$

Thus, the transformation of coordinates (46) changes the form of the differential equation (45) only to the extent of replacing the coefficient matrix A by a similar matrix.

Given the system (45), what we seek is a matrix E, similar to A, such that (48) is easy to solve. The simplest situation occurs when we can find a diagonal matrix E, i.e., a matrix with nonzero elements along the diagonal only. This corresponds to a complete uncoupling of the system. When this happens the general solution of (48) may be written down immediately. The entire process reduces to considerations of the eigenvalues of the matrix A. By definition, $\lambda \neq 0$ is an eigenvalue of A if λ satisfies the nth degree algebraic equation

$$|A - \lambda I| = 0 \tag{50}$$

This is, in fact, the characteristic equation of the system (45) and we shall refer to its nonzero solutions, not as eigenvalues, but as *characteristic values* or numbers. The zero solutions of (50) are also included as characteristic numbers, and so counting multiplicity, there are exactly n complex characteristic numbers. Using (49), it is easily shown that every matrix E which is similar to A possesses exactly the same characteristic numbers. They are numbers characteristic of the differential system and not of the particular coordinate system used in expressing the dynamics of that system. There accompanies the eigenvalue problem an eigenvector problem. In fact, (50) depicts those values of λ such that nontrivial solutions $\bar{u}$ of the algebraic equation

$$A\bar{u} = \lambda\bar{u} \tag{51}$$

exist. This is *not* the analogue of the eigenvector problem discussed in Chap. 1, however. The latter concerns the eigenvectors of the transformation matrix B of the space variables. An eigenvector of B depicts an invariant direction in the complex n-space, under (46), and although an eigenvector of A, i.e., a solution of (51), similarly would depict an invariant direction under a transformation $\bar{u} = A\bar{x}$, the latter geometric transformation we here have simply no use for (see Exercise 5, however).

Thus we study the eigenvalues of A merely to discover the possible forms into which the differential equation may be recast.

If E is the diagonal matrix

$$E = \begin{pmatrix} \lambda_1 & 0 & 0 & \cdots & 0 \\ 0 & \lambda_2 & 0 & \cdots & 0 \\ \multicolumn{5}{c}{\dotfill} \\ 0 & 0 & \cdots & 0 & \lambda_n \end{pmatrix} \tag{52}$$

then clearly the diagonal elements λ_1, λ_2, . . . , λ_n are the characteristic numbers of E. It is not difficult to show that an n-by-n matrix A with n *distinct* characteristic numbers λ_1, λ_2, . . . , λ_n is always similar to a diagonal matrix. From this, we have the important Theorem 2.

Theorem 2

If the n characteristic numbers of the system (45) are distinct, then there exists a nonsingular matrix B such that

$$BAB^{-1} = \begin{pmatrix} \lambda_1 & 0 & 0 & \cdots & 0 \\ 0 & \lambda_2 & 0 & \cdots & 0 \\ \multicolumn{5}{c}{\dotfill} \\ 0 & 0 & \cdots & 0 & \lambda_n \end{pmatrix}$$

The principal matrix solution of (48) is then

$$U = \begin{pmatrix} e^{\lambda_1 t} & 0 & 0 & \cdots & 0 \\ 0 & e^{\lambda_2 t} & 0 & \cdots & 0 \\ \multicolumn{5}{c}{\dotfill} \\ 0 & 0 & 0 & \cdots & e^{\lambda_n t} \end{pmatrix}$$

and the columns of $B^{-1}U$ form a base of solutions of (45). In particular, each component of each solution of (45) is a linear combination of the exponentials $e^{\lambda_1 t}$, $e^{\lambda_2 t}$, . . . , $e^{\lambda_n t}$.

Some of the λ_j in Theorem 2 may be complex. In such a case, real solutions are obtained in the usual fashion by combining terms corresponding to complex conjugates among the λ_j.

EXAMPLE 6

In Case 2, Chap. 1, we have

$$A = \begin{pmatrix} 0 & 1 \\ -\omega^2 & -k \end{pmatrix}$$

and the characteristic numbers satisfy the equation

$$|A - \lambda I| = \begin{vmatrix} -\lambda & 1 \\ -\omega^2 & -k - \lambda \end{vmatrix} = \lambda^2 + k\lambda + \omega^2 = 0$$

As a matter of fact, in this case, we have obtained the transformation matrix B explicitly [see Eqs. (22) of Chap. 1].

$$B = \begin{pmatrix} \omega^2 & -\lambda_1 \\ \omega^2 & -\lambda_2 \end{pmatrix}$$

The $\bar{u}$ equation is given by Eqs. (14) of Chap. 1 with

$$\bar{u} = \begin{pmatrix} u \\ v \end{pmatrix}$$

The construction of the transformation matrix B in this two-dimensional case can be generalized to the n-dimensional case to yield a proof of the diagonalization theorem (Theorem 2) in general.

* * * * * * *

In exceptional cases, Eq. (50) has repeated roots. In such a case, A is, in general, not similar to a diagonal matrix. On the other hand, for the purpose of solving the differential equations it is not necessary to uncouple the system (45) completely. If we can uncouple it in one direction (up or down) then the solutions are readily obtained. To this end we again turn to a theorem of matrix theory.

THEOREM 3

Given an n-by-n matrix A, there exists a nonsingular matrix B such that $E = BAB^{-1}$ is a triangular matrix of the form

$$E = \begin{pmatrix} \lambda_1 & 0 & 0 & \cdots & 0 & 0 \\ \beta_{21} & \lambda_2 & 0 & \cdots & 0 & 0 \\ \beta_{31} & \beta_{32} & \lambda_3 & \cdots & 0 & 0 \\ \cdot & \cdot & \cdot & \cdot & \cdot & \cdot \\ \beta_{n1} & \beta_{n2} & \cdots & & \beta_{n(n-1)} & \lambda_n \end{pmatrix} \tag{53}$$

Clearly, the characteristic numbers of E (and hence of A) are the entries along the diagonal of (53). Some may be repeated. The system (48) becomes

$$\frac{du_1}{dt} = \lambda_1 u_1$$

$$\frac{du_2}{dt} = \beta_{21} u_1 + \lambda_2 u_2$$

$$\frac{du_3}{dt} = \beta_{31} u_1 + \beta_{32} u_2 + \lambda_3 u_3 \tag{54}$$

$$\cdot \cdot \cdot \cdot \cdot \cdot \cdot \cdot \cdot \cdot \cdot \cdot \cdot \cdot$$

$$\frac{du_n}{dt} = \beta_{n1} u_1 + \beta_{n2} u_2 + \cdots + \beta_{n(n-1)} u_{(n-1)} + \lambda_n u_n$$

Integrating the first of these, we have for u_1

$$u_1 = c_1 e^{\lambda_1 t}$$

with c_1 arbitrary. Substituting this into the second equation of (54), we have

$$\frac{du_2}{dt} = c_1 \beta_{21} e^{\lambda_1 t} + \lambda_2 u_2$$

or, what is the same,

$$\frac{du_2}{dt} - \lambda_2 u_2 = c_1 \beta_{21} e^{\lambda_1 t} \tag{55}$$

Equation (55) is a linear equation whose general solution is

$$u_2 = \frac{c_1 \beta_{21}}{\lambda_1 - \lambda_2} \beta_{21} e^{\lambda_1 t} + c_2 e^{\lambda_2 t} = c_1' e^{\lambda_1 t} + c_2 e^{\lambda_2 t} \tag{56}$$

if $\lambda_1 \neq \lambda_2$, or

$$u_2 = c_1 \beta_{21} t e^{\lambda_1 t} + c_2 e^{\lambda_1 t} = c_1'' t e^{\lambda_1 t} + c_2 e^{\lambda_1 t} \tag{57}$$

if $\lambda_1 = \lambda_2$, with c_2 arbitrary. In a similar way, the third equation of (54) is solved using either (56) or (57). The result is a linear combination of the three exponentials $e^{\lambda_1 t}$, $e^{\lambda_2 t}$, and $e^{\lambda_3 t}$ or, if roots are repeated, of appropriate t multiples of the distinct exponentials.   In general, each $\bar{u}$ component consists of a linear combination of the exponentials $e^{\lambda_1 t}$, $e^{\lambda_2 t}$, . . . , $e^{\lambda_n t}$, corresponding to the n characteristic numbers, or of a linear combination of the distinct exponentials with polynomial coefficients in t.   The maximum degree of any polynomial coefficient is one less than the multiplicity of the corresponding characteristic number.   Clearly, the solutions of Eq. (45) [given through (47)] are of the same form.   Hence, we have Theorem 4.

Theorem 4

Each component of each solution of the nth-order system (45) is a linear combination of the exponentials $e^{\lambda_1 t}$, $e^{\lambda_2 t}$, . . . , $e^{\lambda_k t}$, corresponding to the $k \leq n$ distinct characteristic numbers $\lambda_1, \lambda_2, \ldots, \lambda_k$ of the matrix A, with polynomial coefficients in t. The maximum degree of any polynomial coefficient is one less than the multiplicity of the corresponding characteristic number. In particular, if A has n distinct characteristic numbers, then each component of each solution of (45) is a linear combination of the exponentials $e^{\lambda_1 t}$, $e^{\lambda_2 t}$, . . . , $e^{\lambda_n t}$ with constant coefficients.

Example 7

In Case 3 of Chap. 1 we were concerned with the second-order equation

$$\frac{d^2 x}{dt^2} + k \frac{dx}{dt} + \left(\frac{k}{2}\right)^2 x = 0$$

whose characteristic numbers λ_1 and λ_2 were both equal to $-k/2$.   The transformation matrix B implied by Eq. (24) of Chap. 1 is

$$B = \begin{pmatrix} \left(\dfrac{k}{2}\right)^2 & \dfrac{k}{2} \\ \dfrac{k}{2} & 0 \end{pmatrix}$$

When applied to the vector equivalent of the second-order equation, the system became

$$\frac{du}{dt} = -\frac{k}{2}u$$

$$\frac{dv}{dt} = u - \frac{k}{2}v$$

(see Eqs. (25) of Chap. 1).   This system is of the form (54) with $\lambda_1 = \lambda_2 = -k/2$ and $\beta_{21} = 1$.   The process used in this two-dimensional case to uncouple one equation from the other can be generalized to yield a proof of the triangularization theorem (Theorem 3), in general.

EXERCISES

1. Show that the inverse of a matrix $B = (b_{ij})$ is given by $(\beta_{ij}/|B|)$, where β_{ij} is the cofactor of b_{ji}.

2. Using Exercise 1, show that if the principal matrix solution of (38) is bounded, as $t \to \infty$, and if $\int_0^t \text{trace } A(s)\, ds > m$ for some number m and all $t \geq 0$, then the inverse of the principal matrix solution is bounded as $t \to \infty$.

3. Apply the result in Exercise 2 to the special cases of constant A in (45) and the nth-order equation (23).

4. Show that matrices A and $E = BAB^{-1}$, where B is non-singular, possess the same characteristic numbers.   In fact, show that the characteristic polynomials $|A - \lambda I|$ and $|E - \lambda I|$ are identical.

5. Prove the diagonalization theorem (Theorem 2).   Show that eigenvectors of a matrix A, i.e., nontrivial solutions of (51), are linearly independent if they belong to distinct eigenvalues (characteristic numbers).   Discuss this result as a purely algebraic question and, in turn, as a purely geometric question. Then show that if an n-by-n matrix A has n linearly independent eigenvectors, these vectors form a base for complex n-space. Hence, regarding A as a *transformation* matrix relative to the

coordinate system $\bar{x}$, show that relative to the eigencoordinates $\bar{u} = B\bar{x}$, the equivalent transformation matrix becomes BAB^{-1} and is necessarily a diagonal matrix. Finally, extend the result to a singular matrix A with n distinct characteristic numbers. Note that $A - \epsilon I$ is nonsingular for sufficiently small $|\epsilon|$.

6. It is clear that for a diagonal matrix, the trace of the matrix is equal to the sum of its characteristic numbers. Prove this for the general case by considering the factorization of the characteristic polynomial.

7. Using Exercises 4 and 6, explain why trace A = trace E if A and E are similar matrices. Use this result and Theorem 1 to show that the determinants of the principal matrix solutions of (45) and (48) are identical.

8. Explain why U in Theorem 2 is the principal matrix solution of (48) and why the components of solutions of (45) are linear combinations of the exponentials $e^{\lambda_1 t}$, $e^{\lambda_2 t}$, . . . , $e^{\lambda_n t}$.

9. Prove the triangularization theorem (Theorem 3). Show that there is a base for complex n-space such that the matrix equivalent of A relative to this base is of the form (53). The geometric concept of an underlying transformation (irrespective of coordinate representation) proves helpful. Proceed via mathematical induction on the dimension n of the matrix A. Show that if any eigenvector of the transformation is chosen as the first basic vector, then the triangularization process for A is reduced to that for a matrix of dimension $n - 1$.

10. Using Theorem 4, discuss the behavior of a solution of (45), as $t \to \infty$, in terms of the characteristic numbers of the coefficient matrix A. Note, in particular, the degenerate cases wherein one (or more) of the characteristic numbers is (are) zero. Note also the cases wherein multiple pure imaginary characteristic numbers occur. Illustrate the several possibilities by constructing appropriate examples.

11. Let λ be a root of the nth-degree algebraic equation $f(\lambda) = \lambda^n + b_n \lambda^{n-1} + \cdots + b_1 = 0$ of multiplicity $k > 1$.

Show that each of $x = t^i e^{\lambda t}$, $i = 1, \ldots, (k-1)$ is a solution of the nth-order differential equation

$$\frac{d^n x}{dt^n} + b_n \frac{d^{n-1}x}{dt^{n-1}} + \cdots + b_1 x = 0$$

Note that each of the derivatives $f^i(\lambda)$, $i = 1, \ldots, (k-1)$ vanishes if λ is a root of f of multiplicity k. Construct a base for all solutions of the differential equation. Then construct a fundamental matrix solution of the associated matrix differential equation.

12. Show that $\bar{x} = \bar{c}e^{\lambda t}$ is a solution of the system $d\bar{x}/dt = A\bar{x}$ (A constant) if λ is an eigenvalue of A and $\bar{c}$ a corresponding eigenvector. Discuss the connection between this result and Theorem 2.

4. Linear Systems with Periodic Coefficients

We are concerned here with one very general theorem which characterizes the *form* of the solutions of linear periodic systems.

THEOREM 5

Let $Y(t)$ be the principal matrix solution of the n-by-n-matrix equation

$$\frac{dX}{dt} = P(t)\,X \tag{58}$$

where $P(t)$ is a continuous periodic matrix of period $\tau \neq 0$. Then $Y(t)$ is of the form

$$Y(t) = Q(t)\,e^{Bt} \tag{59}$$

where B is a constant matrix and $Q(t)$ has period τ.

PROOF: First of all, we recall that the symbol e^{Bt} denotes the principal matrix solution of an equation

$$\frac{dX}{dt} = BX \tag{60}$$

where B is a constant matrix. Thus (59) asserts that $Y(t)$ is the product of a periodic matrix and the principal matrix solution of a related linear system with constant coefficients. The matrix $X(t) = Y(t + \tau)$ clearly satisfies (58), and hence,

$$Y(t + \tau) = Y(t) \, C$$

for a suitable constant matrix C. Since $C = Y^{-1}(t) \, Y(t + \tau)$, it is nonsingular and can be expressed in the form

$$C = e^{B\tau} \tag{61}$$

for a suitable matrix B. (This is not an altogether trivial fact. See Exercise 13.) Thus, we have

$$Y(t + \tau) = Y(t) \, e^{B\tau} \tag{62}$$

Letting
$$Q(t) = Y(t) \, e^{-Bt} \tag{63}$$

it follows from (62) that

$$Q(t + \tau) = Y(t + \tau) \, e^{-B(t+\tau)} = Y(t) \, e^{B\tau} e^{-B(\tau+t)} = Y(t) \, e^{-Bt}$$
$$= Q(t)$$

Hence $Q(t)$ has period τ and (63) is merely another expression of (59).

It is of interest to observe that $Q(t)$ satisfies the equation

$$\frac{dQ}{dt} = P(t) \, Q - QB \tag{64}$$

Thus B is a matrix such that the above matrix equation possesses a periodic solution of period τ. But this matrix equation is equivalent to a linear system of order n^2 with periodic coefficients of period τ. In vector form, the latter might be written as $d\bar{q}/dt = P^*(t) \, \bar{q}$, where $P^*(t)$ is periodic with period τ. This system of order n^2 is to possess a periodic solution of period τ. In general, there will be numerous choices for B. This is clear from (61) since C does not determine B uniquely. The "exponent" $B\tau$, as it were, is one of the complex "logarithms" of C.

EXAMPLE 8

In the scalar case, Eq. (58) becomes $dx/dt = p(t) x$ with general solution $x(t) = c \exp \left[\int_0^t p(s) \, ds \right]$ for arbitrary c.   Equation (64) becomes $dq/dt = [p(t) - b]q$ with general solution

$$q(t) = q_0 \exp \left[\int_0^t [p(s) - b] \, ds \right]$$

for arbitrary q_0.

Clearly, $q(t)$ has period τ if $\int_0^\tau [p(s) - b] \, ds = 0$, or what is the same, if $b = \frac{1}{\tau} \int_0^\tau p(s) \, ds$.   To the latter one may add any integral multiple of $2\pi i$.

EXAMPLE 9

Consider the linear equation (Hill's equation)

$$\frac{d^2x}{dt^2} + p_1(t) \frac{dx}{dt} + p_2(t) \, x = 0 \tag{65}$$

in which the coefficients $p_1(t)$ and $p_2(t)$ are continuous and have a common period τ.   A nontrivial solution $x(t)$ of (65) is called a *normal solution* if

$$x(t + \tau) = \sigma x(t) \tag{66}$$

holds identically in t for a suitable (complex) constant σ.   Typically, there exist two values of σ (called *characteristic multipliers*) for which there are corresponding normal solutions.   In fact, if $x_1(t)$ and $x_2(t)$ are any given base of real solutions of (65), then each of $x_1(t + \tau)$ and $x_2(t + \tau)$ is a solution of (65) and so

$$\begin{aligned} x_1(t + \tau) &= a_{11}x_1(t) + a_{12}x_2(t) \\ x_2(t + \tau) &= a_{21}x_1(t) + a_{22}x_2(t) \end{aligned} \tag{67}$$

for suitable real constants a_{11}, a_{12}, a_{21}, and a_{22}, not all zero.   Any normal solution $x(t)$ may be expressed in the form

$$x(t) = \alpha_1 x_1(t) + \alpha_2 x_2(t) \tag{68}$$

for suitable complex constants α_1 and α_2 (not both zero).   Thus, if $x(t + \tau) = \sigma x(t)$, (67) and (68) imply that $\sigma\alpha_1 = \alpha_1 a_{11} + \alpha_2 a_{21}$, $\sigma\alpha_2 = \alpha_1 a_{12} + \alpha_2 a_{22}$, or what is the same

$$\alpha_1(a_{11} - \sigma) + \alpha_2 a_{21} = 0$$
$$\alpha_1 a_{12} + \alpha_2(a_{22} - \sigma) = 0$$

If this system is to possess a nontrivial solution (α_1, α_2), then the determinant

$$\begin{vmatrix} a_{11} - \sigma & a_{21} \\ a_{12} & a_{22} - \sigma \end{vmatrix} \tag{69}$$

must vanish.   Thus, σ is necessarily a root of a certain quadratic equation.   On the other hand, if σ is a nonzero root of this quadratic equation, one may retrace the above steps to (66) and therein exhibit a normal solution.   Thus, normal solutions always exist since the quadratic polynomial (69) must process nontrivial roots (see Exercise 4).   When (69) possesses distinct roots, there are two linearly independent, normal solutions, and hence the properties of all solutions are reflected in these two normal solutions.   In particular, for $|\sigma| > 1$, a normal solution satisfies $\lim_{t \to \infty} |x(t)| = \infty$, and for $|\sigma| < 1$, $\lim_{t \to \infty} |x(t)| = 0$.   Only for $\sigma = 1$, of course, is $x(t)$ periodic with period τ.   For $\sigma = -1$, $x(t)$ is periodic with period 2τ.   Periodic solutions of period τ or 2τ are actually uncommon.

EXAMPLE 10

Consider the special Hill's equation

$$\frac{d^2x}{dt^2} + p(t)\, x = 0 \tag{70}$$

where $p(t)$ is continuous and periodic with period τ.   Then the determinant (69) of the previous example becomes

$$\sigma^2 - a\sigma + 1 \tag{71}$$

with "a" real. In fact, the constant term in (69) is the determinant

$$\begin{vmatrix} a_{11} & a_{21} \\ a_{12} & a_{22} \end{vmatrix} \tag{72}$$

which, in turn, is equal to the determinant of the transformation (67). On the other hand by Eq. (34), the Wronskian of any two solutions of (70) is constant; i.e., the first derivative does not appear in (70) and so trace $A = 0$, and since the Wronskian of the pair $x_1(t + \tau)$, $x_2(t + \tau)$ is the product of the determinant (72) and the Wronskian of the pair $x_1(t)$, $x_2(t)$, necessarily

$$\begin{vmatrix} a_{11} & a_{21} \\ a_{12} & a_{22} \end{vmatrix} = 1$$

Now the roots σ_1 and σ_2 of (71) satisfy the relation $\sigma_1\sigma_2 = 1$. Thus, some solutions of (70) are unbounded whenever one of $|\sigma_1|$ and $|\sigma_2|$ is greater than 1 or, what is the same, less than 1. Only in the case

$$\begin{aligned} |\sigma_1| &= 1 \\ |\sigma_2| &= 1 \end{aligned} \tag{73}$$

is it possible for *all* solutions of (70) to be bounded. Since the roots are given by

$$\sigma_1 = \frac{a}{2} - \sqrt{\left(\frac{a}{2}\right)^2 - 1}, \qquad \sigma_2 = \frac{a}{2} + \sqrt{\left(\frac{a}{2}\right)^2 - 1} \tag{74}$$

(73) occurs if and only if $(a/2)^2 \geq 1$. Generally, if $\sigma_1 = \sigma_2 = 1$, or $\sigma_1 = \sigma_2 = -1$, then some solutions are still unbounded. Thus, typically all solutions of (70) are bounded if and only if the characteristic multipliers are nonreal. It is of interest to note that the special circumstances $a/2 = \sigma_1 = \sigma_2 = 1$ or $\sigma_1 = \sigma_2 = -1$ are the only cases wherein periodic solutions of period τ or 2τ exist. Thus, those cases wherein such periodic solutions exist correspond to boundaries between stable and unstable phenomena of (70). In the case of the Mathieu equation,

$d^2x/dt^2 + (\omega^2 + \epsilon \cos t)x = 0$, only for very special pairs of values of (ω^2, ϵ) do periodic solutions with period 2π or 4π exist.

EXAMPLE 11

Actually, the situation depicted in Example 10 is not too special. For although Eq. (70) appears as a special case of the more general Hill's equation (65), the solutions of the latter are reflected in a related equation of the form (70). In fact, the substitution

$$y(t) = x(t) \exp\left[\tfrac{1}{2} \int_0^t p_1(s) \, ds \right] \tag{75}$$

transforms Eq. (65) to

$$\frac{d^2y}{dt^2} + \left(p_2 - \frac{1}{2}\frac{dp_1}{dt} - \frac{1}{4}p_1^2 \right) y = 0 \tag{76}$$

provided $p_1(t)$ is continuously differentiable. But if $p_1(t)$ and $p_2(t)$ have period τ, then the coefficient of y in (76) also has period τ and (76) is of the special form (70).

If (75) is applied to the equation

$$\frac{d^2x}{dt^2} + k\frac{dx}{dt} + (\omega^2 + \epsilon \cos t)\, x = 0 \tag{77}$$

where k is a constant, there results the equation

$$\frac{d^2y}{dt^2} + \left(\omega^2 - \frac{k^2}{4} + \epsilon \cos t \right) y = 0$$

which is the Mathieu equation with ω^2 replaced by $\omega^2 - (k/2)^2$. In this case, $x(t) = y(t)\, e^{-kt/2}$ so that for small k, the effect of the damping term in (77) is quantitatively similar to the effect of a corresponding damping term for an equation with constant coefficients.

EXERCISES

1. With reference to Theorem 5, explain why the principal matrix solution satisfies $Y(t + \tau) = Y(t)\, C$ for a suitable constant matrix C.

2. Regarding $Q(t)$ in Theorem 5 as a transformation matrix, explain why the system (60) is equivalent to the system (58).

3. Discuss the form of the solutions of the vector equation $d\bar{x}/dt = P(t)\ \bar{x}$ as implied by Theorems 4 and 5. In particular, note the form of the solutions when the matrix B of Eq. (60) is, or is equivalent to, a diagonal matrix.

4. Give a justification for the "expansions" in (67) and (68). Explain the derivation of the quadratic polynomial (69) and show that $\sigma = 0$ is never a root.

5. Show how to exhibit a normal solution of (65) which corresponds to a root of the quadratic in (69).

6. Show that normal solutions of (65), corresponding to two distinct characteristic multipliers, are linearly independent.

7. Show that each normal solution $x(t)$ of (65) may be expressed in the form $x(t) = q(t)\ e^{\delta t}$, with $q(t)$ periodic of period τ and δ a suitable (complex) constant. The latter is called a *characteristic exponent* of (65). Discuss the relationship between this form of solution and that implied by Theorems 4 and 5.

8. Discuss the nature of the solutions of (65) as reflected in its characteristic exponents. In particular, discuss the solutions of the special Hill's equation (70).

9. Show how to construct real solutions from nonreal normal solutions of (65).

10. In connection with Example 10, explain why the Wronskian of the pair $x_1(t + \tau)$, $x_2(t + \tau)$ is the product of the determinant (72) and the Wronskian of the pair $x_1(t)$, $x_2(t)$. Explain why the Wronskians are not zero and the significance of this fact in the discussion of (72).

11. Derive (64) and (76).

12. Apply (75) to the equation

$$\frac{d^2x}{dt^2} + 2\epsilon \sin t \frac{dx}{dt} + (\omega^2 + \epsilon \cos t)x = 0$$

and express the result as a Mathieu equation.

13. Show that if C is a nonsingular matrix and τ is a real number different from zero, then there exists a matrix B such that $C = e^{B\tau}$. [This is to justify (61).]   First, note that according to Theorem 3, one may assume C is a triangular matrix.   For if $C = e^{B\tau}$, then $TCT^{-1} = Te^{B\tau}T^{-1} = e^{\tau TBT^{-1}}$.   (Prove this result and explain the significance.)   Observe that

$$\begin{pmatrix} b_1 & 0 \\ b_2 & b_3 \end{pmatrix}^2 = \begin{pmatrix} b_1{}^2 & 0 \\ b_2(b_1 + b_3) & b_3{}^2 \end{pmatrix}$$

and, more generally,

$$\begin{pmatrix} b_1 & 0 \\ b_2 & b_3 \end{pmatrix}^k = \begin{pmatrix} b_1{}^k & 0 \\ b_2 P_k & b_3{}^k \end{pmatrix} \qquad k = 2, 3, \ldots$$

where $P_k = b_1{}^{k-1} + b_1{}^{k-2}b_3 + \cdots + b_1 b_3{}^{k-2} + b_3{}^{k-1}$.   Then explain how to choose the matrix

$$B = \begin{pmatrix} b_1 & 0 \\ b_2 & b_3 \end{pmatrix}$$

so that $$e^{B\tau} = \sum_{k=0}^{\infty} \frac{B^k \tau^k}{k!} = C = \begin{pmatrix} c_1 & 0 \\ c_2 & c_3 \end{pmatrix}$$

for given c_1, c_2, and c_3 with $c_1 \neq 0$ and $c_3 \neq 0$.   Explain why this proves the desired result for $n = 2$.   Finally, extend this technique to a proof (via mathematical induction) for the general case.[1]

14. Let $x(t)$ be a normal solution of (70) which does not vanish for any t.   Then (70) may be expressed in the form

$$\frac{(d^2x/dt^2)}{x} + p(t) = 0$$

Use this form to show that $\int_0^\tau p(t)\, dt = -\int_0^\tau [(dx/dt)^2/x^2]\, dt$, where τ is the relevant period of $p(t)$.   Hence if $\int_0^\tau p(t)\, dt > 0$,

---

[1] Richard Bellman, "Stability Theory of Differential Equations," pp. 29–31, McGraw-Hill Book Company, Inc., New York, 1953.

necessarily $x(t)$ is a nonreal normal solution. Explain the significance of the latter fact.

15. Let $x(t)$ be a real normal solution of (70) which vanishes at t_1 and t_2, with $0 \le t_1 < t_2 \le \tau$, but is positive on the interval $t_1 < t < t_2$. Let $x(t)$ assume its maximum for this interval at t_3. Then explain why

$$\frac{1}{t_3 - t_1} + \frac{1}{t_2 - t_3} = \left[\frac{x(t_3) - x(t_1)}{t_3 - t_1} - \frac{x(t_2) - x(t_3)}{t_2 - t_3} \right] \frac{1}{x(t_3)}$$

$$= \frac{\left(\dfrac{dx}{dt}\right)_1 - \left(\dfrac{dx}{dt}\right)_2}{x(t_3)}$$

where $(dx/dt)_1$ and $(dx/dt)_2$ denote values of the derivative dx/dt for suitable t in the interval $t_1 < t < t_2$. Further, explain why

$$\frac{4}{\tau} \le \frac{4}{t_2 - t_1} \le \frac{1}{t_3 - t_1} + \frac{1}{t_2 - t_3}$$

and why

$$\frac{\left(\dfrac{dx}{dt}\right)_1 - \left(\dfrac{dx}{dt}\right)_2}{x(t_3)} < \int_{t_1}^{t_2} \frac{\left|\dfrac{d^2x}{dt^2}\right|}{|x(t)|} \, dt \le \int_0^\tau \frac{\left|\dfrac{d^2x}{dt^2}\right|}{|x(t)|} \, dt$$

Thus, show that if $\int_0^\tau |p(t)| \, dt \le 4/\tau$ and if $x(t)$ is a normal solution of (70) which vanishes for some t, then necessarily $x(t)$ is a nonreal normal solution.

16. Using Exercises 14 and 15 and a result derived in Example 10, show that all solutions of (70) are bounded as $t \to \infty$ provided $\int_0^\tau p(t) \, dt \ge 0$ and $\int_0^\tau |p(t)| \, dt \le 4/\tau$. We assume here that $p(t)$ does not vanish identically.

17. Apply the result in Exercise 16 to the Mathieu equation.

18. Use a technique introduced in the first set of exercises of this chapter to construct a "second" (generally nonnormal) solution of (70) when (71) possesses a double root. Explain why this solution typically is not bounded.

19. Let $x_1(t)$ in (67) be a normal solution of (65) corresponding to a characteristic multiplier σ and a period τ of (65). Show that

for $x_1(t) \neq 0$, the ratio $R(t) = x_2(t)/x_1(t)$ satisfies the difference equation

$$R(t + \tau) = \frac{a_{21}}{\sigma} + \frac{a_{22}}{\sigma} R(t)$$

From this, show that for each integer $k = 1, 2, \ldots$, we have

$$R(t + k\tau) = \frac{a_{21}}{\sigma} \left[1 + \frac{a_{22}}{\sigma} + \left(\frac{a_{22}}{\sigma}\right)^2 + \cdots + \left(\frac{a_{22}}{\sigma}\right)^{k-1} \right]$$
$$+ \left(\frac{a_{22}}{\sigma}\right)^k R(t)$$

and that

$$x_2(t + k\tau) = a_{21}(\sigma^{k-1} + a_{22}\sigma^{k-2} + \cdots + a_{22}^{k-1})x_1(t) + a_{22}^k x_2(t)$$

Note that the latter is true also for $x_1(t) = 0$, and hence, quite generally true. Explain why

$$x_2(t + k\tau) = ka_{21}\sigma^{k-1}x_1(t) + \sigma^k x_2(t)$$

if σ is a double root of the quadratic (69) and if, in addition, $x_1(t)$ is periodic, why

$$x_2(t + k\tau) = ka_{21}x_1(t) + x_2(t)$$

Thus, unless $a_{21} = 0$, $x_2(t)$ is not bounded.

5. *Asymptotic Behavior of Solutions of Linear Systems*

Here we are concerned with the ultimate behavior of solutions of a linear equation. In particular, we ask:

a. Are all solutions of the equation

$$\frac{d\bar{x}}{dt} = A(t)\, \bar{x} \tag{78}$$

bounded as $t \to \infty$?

b. Do all solutions of (78) approach zero as $t \to \infty$?

We, of course, assume that $A(t)$ is defined and is continuous for all $t \geq 0$.

Since the general solution of (78) may be expressed in the form

$$\bar{x}(t) = Y(t) \; \bar{c} \tag{79}$$

where $Y(t)$ is the principal matrix solution, questions a and b concern the ultimate behavior of the principal matrix solution itself. Thus, all solutions of (78) are bounded or approach zero as $t \to \infty$ if and only if the principal matrix solution $Y(t)$ is bounded or approaches zero as $t \to \infty$.

Two important cases of (78) were considered in the previous sections. If $A(t)$ is either a constant matrix A^* or a periodic matrix $P(t)$, then the principal matrix solution is of the form $Y(t) = Q(t) \; e^{Bt}$, with either $Q(t)$ constant and $B = A^*$ or $Q(t)$ periodic. Thus, the ultimate behavior of $Y(t)$ is reflected in the set of characteristic numbers of B (or A^*). In particular we have:

 c. If the real part of each characteristic number is negative, $Y(t)$ approaches zero as $t \to \infty$.

 d. If the real part of at least one characteristic number is positive, $Y(t)$ is unbounded as $t \to \infty$.

Any statement concerning mere boundedness of $Y(t)$ would necessarily be less categorical than c. On the other hand, d does not characterize the unbounded case.

Other cases with which we are particularly concerned are those in which (78) is ultimately near one of these two types; that is, $A(t)$ is asymptotic to, or approximately equal to, either a constant matrix A or a periodic matrix $P(t)$ as $t \to \infty$. The results are expressed in a number of comparison theorems, and it will be seen that comparison is generally effective only if the approach to the asymptotic form is sufficiently rapid.

If $\lim\limits_{t \to \infty} A(t) = A^*$, where A^* is a constant matrix, it is natural to expect that as $t \to \infty$, the solutions of (78) would resemble those of the equation

$$\frac{d\bar{x}}{dt} = A^*\bar{x} \tag{80}$$

On the other hand, the perturbation matrix $B(t) = A(t) - A^*$, acting over a large interval of time, can induce large variations in the expected behavior, even though $B(t)$ is small and tends to zero as $t \to \infty$. The question arises as to the meaning of the adjective *small* when applied to $B(t)$ for $t \to \infty$. The matrix $B(t)$ may be thought of as arising from some form of parametric excitation. The quantity $\int_0^t |B(s)| \, ds$, therefore, is an index of the maximum *impulse* imparted to the system through parametric excitation up to time t. If this were unbounded as a function of t, arbitrarily large variations in the responses might certainly be expected. Thus, it is reasonable to define "small parametric excitation" to mean $\int_0^\infty |B(s)| \, ds < \infty$. This is a rather strong form of $\lim_{t \to \infty} B(t) = 0$. In such a case, we shall say that $B(t)$ is *impulsively small* as $t \to \infty$. With this reasonable meaning of "smallness" we have Theorem 6.

THEOREM 6

If all solutions of (80) are bounded as $t \to \infty$, the same is true of (78), provided the difference $B(t) = A(t) - A^*$ is impulsively small as $t \to \infty$. If all solutions of (80) approach zero as $t \to \infty$, the same is true of (78), provided $|B(t)| \leq c$ for sufficiently small (positive) c, depending upon A^*.

PROOF: Equation (78) may be written in the form

$$\frac{d\bar{x}}{dt} = A^*\bar{x} + B(t)\,\bar{x}$$

Regarding $B(t)\,\bar{x}$ as a forcing function, we have from (44),

$$\bar{x}(t) = Y(t)\,\bar{c} + \int_0^t Y(t - s)\,B(s)\,\bar{x}(s)\,ds \tag{81}$$

where $Y(t)$ is the principal matrix solution of the associated matrix equation $dX/dt - A^*X$, and $\bar{c}$ is an initial vector. The hypotheses of the first part of the theorem imply that $Y(t)$ is bounded as $t \to \infty$. Thus, we have

$$\left| \bar{x}(t) \right| \leq c_1 \left| \bar{c} \right| + \int_0^t c_1 \left| B(s) \right| \left| \bar{x}(s) \right| ds$$

where $c_1 = \max_{0 \leq t \leq \infty} \left| Y(t) \right|$. Applying Gronwall's lemma, we have further

$$\left| \bar{x}(t) \right| \leq c_1 \left| \bar{c} \right| \exp\left[c_1 \int_0^t \left| B(s) \right| ds \right]$$

Thus, the boundedness of $\int_0^t \left| B(s) \right| ds$ ensures the boundedness of $\bar{x}(t)$. For a proof of the second part of the theorem, we bring to bear the strong exponential decay property of $Y(t)$ which is implied by the hypotheses. In fact, there exist positive constants α and c_2 such that

$$\left| Y(t) \right| \leq c_2 e^{-\alpha t} \tag{82}$$

for all $t \geq 0$. ($-\alpha$ might be taken as the largest of the real parts of the characteristic numbers of A.) Then from (81) we have

$$\left| \bar{x}(t) \right| \leq \left| \bar{c} \right| c_2 e^{-\alpha t} + \int_0^t c_2 e^{-\alpha(t-s)} \left| B(s) \right| \left| \bar{x}(s) \right| ds$$

or, what is the same,

$$\left[\left| \bar{x}(t) \right| e^{\alpha t} \right] \leq \left| \bar{c} \right| c_2 + \int_0^t c_2 \left| B(s) \right| \left[\left| \bar{x}(s) \right| e^{\alpha s} \right] ds$$

Hence by Gronwall's lemma,

$$\left[\left| \bar{x}(t) \right| e^{\alpha t} \right] \leq \left| \bar{c} \right| c_2 \exp\left[c_2 \int_0^t \left| B(s) \right| ds \right]$$

or, what is the same,

$$\left| \bar{x}(t) \right| \leq \left| \bar{c} \right| c_2 \exp\left[c_2 \int_0^t \left| B(s) \right| ds - \alpha t \right] \tag{83}$$

Thus if $\left| B(t) \right| \leq c < \alpha/c_2$, (83) implies that

$$\left| \bar{x}(t) \right| \leq \left| \bar{c} \right| c_2 e^{(c_2 c - \alpha)t}$$

with $(c_2 c - \alpha) < 0$, and so $\lim_{t \to \infty} \bar{x}(t) = \bar{0}$. Finally, we note that the bound α/c_2 for $|B(t)|$ depends upon the principal matrix solution $Y(t)$ which, in turn, depends upon the constant matrix A^* [see (82)]. Thus the area of application is determined solely by the approximate asymptotic form of the equation.

The second part of Theorem 6 implies that the system can absorb a total infinite impulse provided the system is positively damped and the impulse is imparted to the system at a sufficiently slow rate. Such a result is, of course, to be expected. The expected also occurs in the inhomogeneous case. By similar arguments we can prove Theorem 7.

THEOREM 7

If all solutions of (80) are bounded as $t \to \infty$, then the same is true of the inhomogeneous equation

$$\frac{d\bar{x}}{dt} = A(t)\,\bar{x} + \bar{f}(t) \tag{84}$$

provided each of $B(t) = A(t) - A^*$ and $\bar{f}(t)$ is impulsively small as $t \to \infty$, i.e.,

$$\int_0^\infty |B(t)|\,dt < \infty \qquad \text{and} \qquad \int_0^\infty |\bar{f}(t)|\,dt < \infty$$

If all solutions of (80) approach zero as $t \to \infty$, then all solutions of (84) are bounded as $t \to \infty$ provided $\bar{f}(t)$ is bounded as $t \to \infty$ and $B(t)$ is impulsively small as $t \to \infty$.

PROOF: Using (44), we have, for any solution of (84),

$$\bar{x}(t) = Y(t)\,\bar{c} + \int_0^t Y(t - s)[B(s)\,\bar{x}(s) + \bar{f}(s)]\,ds \tag{85}$$

where $Y(t)$ is the principal matrix solution of $dX/dt = A^*X$, and $\bar{c}$ is an initial vector. The hypotheses of the first part of the theorem imply that

$$|\bar{x}(t)| \le c_1|\bar{c}| + \int_0^t [c_1\,|B(s)|\,|\bar{x}(s)| + c_1\,|\bar{f}(s)|]\,ds$$

for a suitable positive constant c_1. Applying Gronwall's lemma, we have further

$$\mathopen| \bar{x}(t) \mathclose| \leq c_2 \exp \left[c_1 \int_0^t \mathopen| B(s) \mathclose| \, ds + \frac{c_1}{c_2} \int_0^t \mathopen| \tilde{f}(s) \mathclose| \, ds \right]$$

where c_2 is any positive constant, not smaller than $c_1 \mathopen| \bar{c} \mathclose|$. Thus, $\bar{x}(t)$ is bounded as $t \to \infty$ if each of $\int_0^t \mathopen| B(s) \mathclose| \, ds$ and $\int_0^t \mathopen| \tilde{f}(s) \mathclose| \, ds$ is. For the proof of the second part of the theorem, we first consider the asymptotic form of the equation

$$\frac{d\bar{y}}{dt} = A^* \bar{y} + \tilde{f}(t) \tag{86}$$

According to (44),

$$\bar{y}(t) = \int_0^t Y(t - s) \, \tilde{f}(s) \, ds$$

is a solution of (86) and since the hypotheses of the second part of the theorem imply that $\mathopen| Y(t) \mathclose| \leq c_3 e^{-\alpha t}$ and $\mathopen| \tilde{f}(t) \mathclose| \leq c_4$ for suitable positive constants c_3, c_4, and α, we have

$$\mathopen| \bar{y}(t) \mathclose| \leq c_3 c_4 \int_0^t e^{-\alpha(t-s)} \, ds = \frac{c_3 c_4}{\alpha} (1 - e^{-\alpha t})$$

Thus, $\bar{y}(t)$ is certainly bounded as $t \to \infty$. On the other hand, if we subtract (86) from (84) and let $\bar{z} = \bar{x} - \bar{y}$, we obtain for $\bar{z}$, the linear equation

$$\frac{d\bar{z}}{dt} = A(t) \, \bar{z} + B(t) \, \bar{y} \tag{87}$$

with $B(t) \, \bar{y}$ as a forcing function. Since $\bar{y}$ is bounded and $B(t)$ is impulsively small as $t \to \infty$, it follows that the product $B(t) \, \bar{y}$ is impulsively small as $t \to \infty$. Thus, the hypotheses of the first part of the theorem are fulfilled in Eq. (87), and we conclude that every solution $\bar{z}$ is bounded as $t \to \infty$. Clearly then, every solution $\bar{x} = \bar{z} + \bar{y}$ of (84) is bounded as $t \to \infty$.

EXAMPLE 12

All solutions of the equation $d^2x/dt^2 + [1 + 1/(1 + t^2)]x = 0$ are bounded as $t \to \infty$.

EXAMPLE 13

There are unbounded solutions of the equation

$$\frac{d^2x}{dt^2} + (1 + \epsilon \cos t)x = 0$$

for every $|\epsilon|$ sufficiently small.[1]

EXAMPLE 14

If $k > 0$, then all solutions of the equation

$$\frac{d^2x}{dt^2} + \frac{k\,dx}{dt} + \left(1 + \frac{1}{1 + t^2}\right)x = 0$$

tend to zero as $t \to \infty$.

EXAMPLE 15

If $k > 0$, then all solutions of the equation

$$\frac{d^2x}{dt^2} + \frac{k\,dx}{dt} + (1 + \epsilon \cos t)x = 0$$

tend to zero as $t \to \infty$ provided $|\epsilon|$ is sufficiently small.

EXAMPLE 16

All solutions of the equation $d^2x/dt^2 + [1 + 1/(1 + t^4)]x = e^{-t}$ are bounded as $t \to \infty$.

EXAMPLE 17

If $k > 0$, then all solutions of the equation

$$\frac{d^2x}{dt^2} + \frac{k\,dx}{dt} + \left(1 + \frac{1}{1 + t^4}\right)x = \cos t$$

are bounded as $t \to \infty$.

[1] N. W. McLachlan, "Theory and Application of Mathieu Functions," Oxford University Press, New York, 1947.

EXAMPLE 18

There are no bounded solutions of the equation

$$\frac{d^2x}{dt^2} + \left(1 + \frac{1}{1+t^2}\right)x = \cos t$$

(see Exercise 4).

EXERCISES

1. Explain the statements in c and d and the relevance of c to the inequality (82).

2. Discuss the significance of each of the Examples 12 to 18 in relation to Theorems 6 and 7. In particular, indicate those instances of direct application of the theorems and those instances where direct application is not justified. For the latter, explain why the theorems are not germane.

3. Introduce the change of variable (75) in each of Examples 14 and 15, and explain why these cases are (or are not) covered in Examples 12 and 13.

4. Using the technique introduced in the proof of Theorem 7, show that there are no bounded solutions x of the equation $d^2x/dt^2 + [1 + 1/(1 + t^4)]x = \cos t$. In fact, show that if y is any solution of the asymptotic equation $d^2y/dt^2 + y = \cos t$, then the difference $z = x - y$ satisfies a linear equation which admits only bounded solutions. Explain why this proves the desired result.

5. Show that for $t > 0$, $x = t \sin t$ satisfies the linear equation $d^2x/dt^2 - (1/t)\,dx/dt + (1 + 1/t^2)x = 0$. Explain the significance of such an example.

6. Introduce the change of variable (75) in the equation $d^2x/dt^2 + (1/t)\,dx/dt + x = 0$, and then use Theorem 6 to show that every nontrivial solution for $t > 0$ tends to zero as $t \to \infty$. Note that the decay is not of the exponential type, however.

7. Show that each nontrivial solution of the equation

$$\frac{d^2x}{dt^2} - (1/t)\frac{dx}{dt} + x = 0$$

for $t > 0$ is unbounded as $t \to \infty$.

$\bullet$ $\bullet$ $\bullet$ $\bullet$ $\bullet$ $\bullet$ $\bullet$

If in (78), $A(t)$ is asymptotic to a periodic matrix $P(t)$ as $t \to \infty$, we use the representation theorem, Theorem 5, to prove Theorem 8.

THEOREM 8

If all solutions of the periodic system

$$\frac{d\bar{x}}{dt} = P(t)\,\bar{x} \tag{88}$$

are bounded as $t \to \infty$, then the same is true for (78), provided the difference $C(t) = A(t) - P(t)$ is impulsively small as $t \to \infty$. If all solutions of (78) approach zero as $t \to \infty$, then the same is true of (88) provided $\|C(t)\| \leq c$ for sufficiently small (positive) c, depending upon the approximate asymptotic form (88).

PROOF: Let $Y(t) = Q(t)\,e^{Bt}$ be the principal matrix solution of $dX/dt = P(t)\,X$, where $Q(t)$ is periodic. Then (78) may be expressed in the form

$$\frac{d\bar{x}}{dt} = P(t)\,\bar{x} + C(t)\,\bar{x}$$

and for any solution, we have from (43),

$$\bar{x}(t) = Y(t)\,\bar{c} + \int_0^t [Y(t)\,Y^{-1}(s)\,C(s)\,\bar{x}(s)]\,ds$$

$$= Y(t)\,\bar{c} + \int_0^t [Q(t)\,e^{Bt}e^{-Bs}Q^{-1}(s)\,C(s)\,\bar{x}(s)]\,ds$$

where $\bar{c}$ is an initial vector. Hence

$$|\bar{x}(t)| \leq |Y(t)\,\bar{c}| + \int_0^t |Q(t)|\,\|e^{B(t-s)}\|\,\|Q^{-1}(s)\|\,\|C(s)\|\,|\bar{x}(s)|\,ds \tag{89}$$

Now, the inverse of the periodic matrix $Q(t) = (q_{ij}(t))$ has for its ijth element, the quantity $p_{ij}(t)/|Q(t)|$, where $p_{ij}(t)$ is the cofactor of $q_{ji}(t)$. Further, since the determinant $|Q(t)|$ is periodic and does not vanish, it is uniformly bounded away from zero. Thus, along with $Q(t)$, the inverse $Q^{-1}(t)$ is uniformly bounded for $t \geq 0$ and the inequality (89) may be replaced by

$$|\bar{x}(t)| \leq |Y(t)\,\bar{c}| + c_2 \int_0^t |e^{B(t-s)}|\,||C(s)||\,|\bar{x}(s)|\,ds \quad (90)$$

for a suitable positive constant c_2. If all solutions of (88) are bounded, then (90) in turn yields

$$|\bar{x}(t)| \leq c_1|\bar{c}| + c_1 c_2 \int_0^t |C(s)|\,||\bar{x}(s)|\,ds$$

for a suitable positive constant c_1. Thus, by Gronwall's lemma, we have

$$|\bar{x}(t)| \leq c_1|\bar{c}|\exp\left[c_1 c_2 \int_0^t |C(s)|\,ds\right]$$

and the condition $\int_0^\infty |C(s)|\,ds < \infty$ implies $\bar{x}(t)$ is bounded as $t \to \infty$. On the other hand, if all solutions of (88) approach zero as $t \to \infty$, then there exist positive constants c_3, c_4, and α such that

$$|e^{Bt}| \leq c_3 e^{-\alpha t} \qquad \text{and} \qquad |Y(t)| \leq c_4 e^{-\alpha t}$$

Thus, (90) implies that

$$|\bar{x}(t)| \leq c_4|\bar{c}|e^{-\alpha t} + c_2 c_3 \int_0^t e^{-\alpha(t-s)}|C(s)||\bar{x}(s)|\,ds$$

or, what is the same,

$$[|\bar{x}(t)|e^{\alpha t}] \leq c_4|\bar{c}| + c_2 c_3 \int_0^t |C(s)|[|\bar{x}(s)|e^{\alpha s}]\,ds$$

By Gronwall's lemma, therefore, we have

$$[|\bar{x}(t)|e^{\alpha t}] \leq c_4|\bar{c}|\exp\left[c_2 c_3 \int_0^t |C(s)|\,ds\right]$$

or, what is the same,

$$\mid \bar{x}(t) \mid \leq c_4 \mid \bar{c} \mid \exp \left[c_2 c_3 \int_0^t \mid C(s) \mid ds - \alpha t \right]$$

Thus, if $\mid C(s) \mid \leq c < \alpha/(c_2 c_3)$, then $\lim_{t \to \infty} \bar{x}(t) = \bar{0}$.   We note that each of the positive constants c_2, c_3, and α depends solely upon the principal matrix solution of the approximate asymptotic (periodic) form (88).

EXAMPLE 19

If $k > 0$, then all solutions of

$$\frac{d^2x}{dt^2} + k\frac{dx}{dt} + [\omega^2 + p(t)]x = 0$$

approach zero as $t \to \infty$ provided $\mid p(t) - \epsilon \cos t \mid \leq c$ for sufficiently small c and $\mid \epsilon \mid$.   Note that $p(t)$ need not be periodic, but if it is, $\epsilon \cos t$ might be the fundamental in a Fourier expansion of $p(t)$.

✦ ✦ ✦ ✦ ✦ ✦ ✦

In view of Theorems 6 and 8 one might anticipate some rather general comparison theorems.   For example, if all solutions of the equation

$$\frac{d\bar{x}}{dt} = A(t)\,\bar{x} \tag{91}$$

are bounded as $t \to \infty$, then does it follow that all solutions of

$$\frac{d\bar{x}}{dt} = [A(t) + B(t)]\bar{x} \tag{92}$$

are bounded as $t \to \infty$ provided $\int_0^\infty \mid B(t) \mid dt < \infty$?   Such is not the case, as may be seen by examples.   One cannot even conclude in general that all solutions of (92) are bounded if all solutions of (91) approach zero as $t \to \infty$.[1]   However, if the trace

[1] Bellman, *op. cit.*, pp. 42–43.

of the coefficient matrix $A(t)$ is sufficiently well behaved as $t \to \infty$, then general comparison theorems of the above type are available. The following is an important illustration of one such.

Theorem 9

If all solutions of (91) are bounded as $t \to \infty$, then all solutions of (92) are bounded as $t \to \infty$ provided $B(t)$ is impulsively small as $t \to \infty$ and $\int_0^t \text{trace } A(s) \, ds$ is bounded away from $-\infty$ as $t \to \infty$.

Proof: Again using (43) we have, for any solution of (92)

$$\bar{x}(t) = Y(t) \, \bar{c} + \int_0^t [Y(t) \, Y^{-1}(s) \, B(s) \, \bar{x}(s)] \, ds$$

where $Y(t)$ is the principal matrix solution of the equation $dX/dt = A(t) \, X$, and $\bar{c}$ is an initial vector. Hence

$$\left| \bar{x}(t) \right| \leq \left| Y(t) \, \bar{c} \right| + \int_0^t \left| Y(t) \right| \left| Y^{-1}(s) \right| \left| B(s) \right| \left| \bar{x}(s) \right| ds$$

The hypotheses imply that $|Y(t)| = \exp \left[\int_0^t \text{trace } A(s) \, ds \right]$ is bounded away from zero, and this, in turn, implies that $\left| Y^{-1}(t) \right|$ is bounded as $t \to \infty$ if $\left| Y(t) \right|$ is. Thus, if all solutions of (91) are bounded as $t \to \infty$, then

$$\left| \bar{x}(t) \right| \leq c_1 \left| \bar{c} \right| + c_2 \int_0^t \left| B(s) \right| \left| \bar{x}(s) \right| ds$$

for suitable $c_1 > 0$ and $c_2 > 0$ and by Gronwall's lemma,

$$\left| \bar{x}(t) \right| \leq c_1 \left| \bar{c} \right| \exp \left[c_2 \int_0^t \left| B(s) \right| ds \right]$$

Hence $\bar{x}(t)$ is bounded as $t \to \infty$ provided $B(t)$ is impulsively small.

Example 20

For the equation $d^2x/dt^2 + a(t) \, x = 0$, trace $A(t) = 0$, so that Theorem 9 in this case is an extension of the first part of Theorem 8 which concerned only the periodic case.

EXAMPLE 21

If $\int_0^\infty |a(t)|\, dt < \infty$, then there are unbounded solutions of the equation $d^2x/dt^2 + a(t)\, x = 0$. In fact, the vector version of this equation $d\bar{x}/dt = A(t)\, \bar{x}$ relates to the matrix

$$A(t) = \begin{pmatrix} 0 & 1 \\ -a(t) & 0 \end{pmatrix}$$

and if all solutions were bounded, Theorem 9 would assert that all solutions of $d\bar{x}/dt = A(t)\, \bar{x} + B(t)\, \bar{x}$, with

$$B(t) = \begin{pmatrix} 0 & 0 \\ a(t) & 0 \end{pmatrix}$$

would be bounded. But this is impossible since the scalar version of the latter is $d^2x/dt^2 = 0$.

◆ ◆ ◆ ◆ ◆ ◆ ◆ ◆

Theorems 8 and 9 (and parts of 6 and 7) are each special cases of the rather general Theorem 10.

THEOREM 10

If all solutions of (91) and its adjoint system $d\bar{z}'/dt = -\bar{z}'\, A(t)$ ($\bar{z}'$ is a row vector) are bounded as $t \to \infty$, then all solutions of (92) are bounded as $t \to \infty$ provided $B(t)$ is impulsively small as $t \to \infty$.

PROOF: A proof, similar to that of Theorem 9, may be based on the observation that all solutions of the adjoint system are bounded as $t \to \infty$ if and only if the inverse of the principal matrix solution $Y(t)$ of $dX/dt = A(t)\, X$ is bounded as $t \to \infty$. For $Y^{-1}(t)$ is the principal matrix solution of the adjoint system.

EXERCISES

1. What theorems of this chapter are required to justify the statements in Example 19?

2. Explain the statements made in Examples 20 and 21.

3. Show that each of Theorem 8 and 9 (and a part of each of 6 and 7) is a special case of Theorem 10.

4. Show that any solution of $d\bar{x}/dt = A(t)\,\bar{x}$ satisfies $\big|\,\bar{x}(t)\,\big| \leq \big|\,\bar{c}\,\big| \exp\Big[\int_0^t \big|\,A(s)\,\big|\,ds\Big]$, where $\bar{c}$ is the initial vector.

5. Show that $\lim_{t \to \infty} \bar{x}(t)$ exists if $d\bar{x}/dt = A(t)\,\bar{x}$ and

$$\int_0^t \big|\,A(s)\,\big|\,ds < \infty.$$

6. Give a detailed proof of Theorem 10.

Chapter 5

STABILITY IN NONLINEAR SYSTEMS

1. *The Concept of Stability*

There are basically three categories of the stability concept: Laplace, Liapunov, and Poincaré. The first is of such generality as to be rarely useful, while the second is so restrictive that one is forced to introduce the third to discuss stability of periodic solutions properly. There are many specialized stability concepts which have been introduced and developed within each of these three basic categories. We shall be concerned with but a few of these.

Laplace stability is a boundedness concept of a very general nature. A system is *stable in the sense of Laplace* if it exhibits only finite motions; i.e., all solutions of the differential equations are bounded as $t \to \infty$. The use of the word "stability" with reference to the stability regions of the Mathieu equation invokes the Laplace concept of stability. Such a stability concept, however, is not useful for quantitative matters in variational problems, since it distinguishes variational effects only as to their being finite or infinite. For example, any earthbound phenomenon is stable in the sense of Laplace.

Liapunov stability, on the other hand, is concerned with very stringent restrictions on the motion. It is required that motions (solutions) which are once near together remain near together

for all future time as functions of the time.   Specifically, we say
that a solution $\bar{x}(t)$ of

$$\frac{d\bar{x}}{dt} = f(\bar{x},t) \tag{1}$$

is *stable in the sense of Liapunov*, if for each $\epsilon > 0$, there exists a
$\delta > 0$ such that any solution $\bar{y}$ of (1) satisfying $|\bar{x} - \bar{y}| \leq \delta$ for
$t = 0$ also satisfies $|\bar{x} - \bar{y}| \leq \epsilon$ for *all* $t \geq 0$.   (The choice of
initial time $t = 0$ in this definition is, of course, of no importance
to the concept.)   Liapunov stability embodies much of what one
desires in a stability concept.   However, the time-dependent
comparison, implied by the inequality $|\bar{x} - \bar{y}| \leq \epsilon$, often pre-
cludes stability for certain steady-state phenomena which *should*
be considered stable.   This difficulty lead Poincaré to the concept of orbital stability.

Orbital stability is concerned with the behavior of positive half
paths of trajectories.   A trajectory Γ is *stable in the sense of
Poincaré*, i.e., possesses *orbital* stability, if neighboring half paths
which are once near Γ remain near Γ.   In this case, "stability"
does not require a comparison between trajectories as functions
of the independent variable.   Geometrically, one thinks of a
trajectory Γ in n-space surrounded by a tubing which has the
property that any trajectory which once penetrates this tubing
must thereafter remain within a slightly larger tubing.   If this
is the case for arbitrarily small tubings, then Γ possesses orbital
stability.   If we further require that there be no "shearing"
within the tubings so that the trajectories move down the
tubings together, then Γ is stable in the more restrictive sense
of Liapunov.

A trajectory Γ is *asymptotically* stable (orbitally or otherwise)
if positive half paths which are once near Γ actually approach
Γ as $t \rightarrow \infty$.

In the following, the words "stability" or "asymptotic stability" will mean stability or asymptotic stability in the sense of

Liapunov and a solution will be *unstable* if it is not stable in the
sense of Liapunov.

2. *Stability of Singular Points of Autonomous Systems*

The solutions of $\bar{f}(\bar{x}) = \bar{0}$ are singular points (point solutions)
of the differential equation

$$\frac{d\bar{x}}{dt} = \bar{f}(\bar{x}) \tag{2}$$

If $\bar{c}$ is such a point, we say that (2) is *stable at* $\bar{c}$ if $\bar{x} = \bar{c}$ is a
stable solution of (2).   We may, on occasion, refer to a singular
point $\bar{c}$ as a stable (or unstable) position of equilibrium.   In particular, the linear system

$$\frac{d\bar{x}}{dt} = A\bar{x} \tag{3}$$

is stable at the origin if $\bar{x} = \bar{0}$ is a stable solution of (3).   From
Chap. 4, we know that (3) is stable at the origin, say, if the real
part of each characteristic number of A is nonpositive and any
pure imaginary characteristic number is at most a simple root
of the characteristic equation.   Equation (3) is asymptotically
stable at the origin if the real part of each characteristic number
of A is actually negative.   More generally, a nonlinear equation
may have several singular points, all, none, or some of which
may be stable.   If $\bar{c}$ is a singular point of (2), we may introduce
a new variable $\bar{y} = \bar{x} - \bar{c}$ and (2) becomes

$$\frac{d\bar{y}}{dt} = \bar{f}(\bar{y} + \bar{c}) = \bar{f}^*(\bar{y})$$

The singularity now appears at the origin $\bar{y} = \bar{0}$ in the new
coordinates.   Thus there is no loss in generality if we assume
that any particular singular point is placed at the origin of the
coordinate system.

EXAMPLE 1

The scalar equation $dx/dt = x - x^2$ possesses two singular points, $x = 0$ and $x = 1$. The latter is asymptotically stable while the former is unstable. With $y = x - 1$, the differential equation becomes $dy/dt = -y - y^2$ so that the stable equilibrium position now appears at the origin, $y = 0$.

EXAMPLE 2

Consider the equation $dy/dt = -y + y^2$. It may be shown that solutions initially near the origin tend to the origin as $t \to \infty$, while solutions initially far from the origin tend to infinity in *finite* time. Thus the origin is asymptotically stable, but distant solutions are not even continuable for all $t \geq 0$.

♦ ♦ ♦ ♦ ♦ ♦ ♦

Suppose $\bar{c}$ is a singular point of (2). We may express the right-hand member of (2) in the form

$$\bar{f}(\bar{x}) = A(\bar{x} - \bar{c}) + \bar{f}^{(1)}(\bar{x}) \tag{4}$$

where A is a (any) constant matrix. Here $\bar{f}^{(1)}(\bar{x})$ is merely the difference $\bar{f}(\bar{x}) - A(\bar{x} - \bar{c})$. On the other hand, suppose $A \neq 0$ can be chosen so that

$$\lim_{\bar{x} \to \bar{c}} \left| \frac{\bar{f}^{(1)}(\bar{x})}{\bar{x} - \bar{c}} \right| = 0 \tag{5}$$

Then the equation

$$\frac{d\bar{x}}{dt} = A(\bar{x} - \bar{c}) \tag{6}$$

deserves the name *linear approximation*. The condition (5) (referred to as the *nonlinearity condition*) is a precise way of saying that the linear term in (4) is the dominant part of $\bar{f}(\bar{x})$ for $\bar{x}$ near $\bar{c}$. Then (4) is "linearized" merely by erasing the second term on the right. For stability considerations, the latter physical process is justified by the mathematical Theorem 1.

THEOREM 1

If the linear approximation (6) is asymptotically stable at $\bar{c}$, then (2) is asymptotically stable at $\bar{c}$.

PROOF: Our proof will imply that solutions of (2) once near $\bar{c}$ remain near $\bar{c}$ and hence exist for all $t \geq 0$. With $\bar{y} = \bar{x} - \bar{c}$, Eq. (2) may be expressed in the form

$$\frac{d\bar{y}}{dt} = A\bar{y} + \bar{f}^{(1)}(\bar{y} + \bar{c}) \tag{7}$$

and the nonlinearity condition (5) may be reformulated as

$$\lim_{\bar{y} \to 0} \frac{|\bar{f}^{(1)}(\bar{y} + \bar{c})|}{|\bar{y}|} - 0 \tag{8}$$

Then from (44) of Chap. 4, we infer that any solution of (7) satisfies an integral equation of the form

$$\bar{y}(t) = Y(t)\,\bar{c}^* + \int_0^t Y(t-s)\,\bar{f}^{(1)}(\bar{y}(s) + \bar{c})\,ds$$

where $Y(t)$ is the principal matrix solution of the associated matrix equation $dX/dt = AX$ and $\bar{c}^*$ is an initial vector. The hypotheses imply that

$$|Y(t)| \leq ce^{-\alpha t} \tag{9}$$

for suitable positive constants c and α. Hence

$$|\bar{y}(t)| \leq c\,|\bar{c}^*|\,e^{-\alpha t} + c\int_0^t e^{-\alpha(t-s)}\,|\bar{f}^{(1)}(\bar{y}(s) + \bar{c})|\,ds$$

or, what is the same,

$$|\bar{y}(t)|\,e^{\alpha t} \leq c\,|\bar{c}^*| + c\int_0^t |\bar{f}^{(1)}(\bar{y}(s) + \bar{c})|\,e^{\alpha s}\,ds \tag{10}$$

But from the nonlinearity condition (8), it follows that there exists a positive number δ such that

$$|\bar{f}^{(1)}(\bar{y} + \bar{c})| \leq \frac{\alpha}{2c}\,|\bar{y}|$$

for $|y| \leq \delta$. Then from (10), we have further

$$[\,|\bar{y}(t)|\,e^{\alpha t}] \leq c\,|\bar{c}^*\,| + \frac{\alpha}{2}\int_0^t [\,|\bar{y}(s)|\,e^{\alpha s}]\,ds \qquad (11)$$

provided $|\bar{y}(s)| \leq \delta$ for $0 \leq s \leq t$. For $t = 0$, we have $|\bar{y}(0)|$ $\leq c\,|\bar{c}^*\,|$ and so if $\bar{c}^*$ is required to satisfy

$$|\bar{c}^*\,| < \delta/c \qquad (12)$$

then the inequality $|\bar{y}(s)| \leq \delta$ will be satisfied on some interval $0 \leq s \leq t$ with $t > 0$. Applying Gronwall's lemma to (11), we have

$$|\bar{y}(t)|\,e^{\alpha t} \leq c\,|\bar{c}^*\,|\,e^{\alpha t/2}$$

or, what is the same,

$$|\bar{y}(t)| \leq c\,|\bar{c}^*\,|\,e^{-\alpha t/2} \qquad (13)$$

But this shows that the inequality $|\bar{y}(t)| \leq c\,|\bar{c}^*\,| < \delta$ once initiated is maintained and thus applies throughout. Clearly, then, (13) applies for all $t \geq 0$ and so $\lim_{t \to \infty} |\bar{y}(t)| = 0$. Finally, we note that the inequality (12) is merely a restriction on the initial perturbation of $\bar{x}(t)$ relative to the singular point $\bar{c}$. Thus the theorem is proved.

The requirement that (6) be *asymptotically* stable at $\bar{c}$ is essential even for proving that the solutions of (2) initially near $\bar{c}$ remain bounded. In general, it is *not* true that the stability of (6) at $\bar{c}$ implies the stability of (2) at $\bar{c}$. The above proof is effective in the asymptotic case because the linear approximation (6) ensures a tendency for perturbation amplitudes to decrease, thereby rendering the influence of the linear term in (4) more and more pronounced. At the same time, the nonlinearity condition ensures that this tendency for perturbation amplitudes to decrease is not disrupted.

EXAMPLE 3

The Duffing equation $d^2x/dt^2 + k\, dx/dt + \omega^2 x + \beta x^3 = 0$ is asymptotically stable at the origin if $k > 0$. In this case we have the system $d\bar{x}/dt = A\bar{x} + \bar{f}^{(1)}(\bar{x})$ where

$$A = \begin{pmatrix} 0 & 1 \\ -\omega^2 & -k \end{pmatrix} \quad \text{and} \quad \bar{f}^{(1)}(\bar{x}) = \begin{pmatrix} 0 \\ -\beta x_1^3 \end{pmatrix}$$

We note that

$$\frac{|\bar{f}^{(1)}(\bar{x})|}{|\bar{x}|} = \frac{|\beta|\,|x_1|^3}{|x_1| + |x_2|} \leq |\beta| x_1^2$$

and so

$$\lim_{\bar{x} \to \bar{0}} \frac{|\bar{f}^{(1)}(\bar{x})|}{|\bar{x}|} = 0$$

EXAMPLE 4

For the equation $d^2x/dt^2 + f(x)\, dx/dt + \omega^2 x = 0$, we let

$$A = \begin{pmatrix} 0 & 1 \\ -\omega^2 & -f(0) \end{pmatrix} \quad \text{and} \quad \bar{f}^{(1)}(\bar{x}) = \begin{pmatrix} 0 \\ [f(0) - f(x_1)]x_2 \end{pmatrix}$$

Then

$$\frac{|\bar{f}^{(1)}(\bar{x})|}{|\bar{x}|} = \frac{|f(0) - f(x_1)|\,|x_2|}{|x_1| + |x_2|} \leq |f(0) - f(x_1)|$$

so that

$$\lim_{\bar{x} \to \bar{0}} \frac{|\bar{f}^{(1)}(\bar{x})|}{|\bar{x}|} = 0$$

if $f(x)$ is continuous. If $f(0) > 0$, the nonlinear equation is asymptotically stable at the origin.

EXAMPLE 5

Consider the stability at the origin of the equation

$$\frac{d^2x}{dt^2} + \frac{dx}{dt} + x - \sqrt{|x|} = 0 \tag{14}$$

Let $A = \begin{pmatrix} 0 & 1 \\ -1 & -1 \end{pmatrix}$, so that $\bar{f}^{(1)}(\bar{x}) = \begin{pmatrix} 0 \\ \sqrt{|x_1|} \end{pmatrix}$. Then the non-linearity condition for $\bar{f}^{(1)}(\bar{x})$ is not satisfied at $\bar{x} = \bar{0}$. Indeed, for $|x_2| = 0$

$$\frac{|\bar{f}^{(1)}(\bar{x})|}{|\bar{x}|} = \frac{\sqrt{|x_1|}}{|x_1| + |x_2|} = \frac{1}{\sqrt{|x_1|}}$$

It is not difficult to show that (14) is unstable at the origin, though that portion of (14) which is obtained by erasing the nonlinear term $\sqrt{|x_1|}$ is, in fact, asymptotically stable at the origin. On the other hand, the equilibrium position $x = 1$ is asymptotically stable. In fact, letting $y = x - 1$, (14) becomes

$$\frac{d^2y}{dt^2} + \frac{dy}{dt} + y + (1 - \sqrt{|y+1|}) = \frac{d^2y}{dt^2} + \frac{dy}{dt} + \frac{1}{2}y$$
$$+ \left(1 - \sqrt{|y+1|} + \frac{1}{2}y\right) = 0 \quad (15)$$

Hence with

$$A = \begin{pmatrix} 0 & 1 \\ -1 & -\frac{1}{2} \end{pmatrix} \quad \text{and} \quad \bar{f}^{(2)}(\bar{y}) = \begin{pmatrix} 0 \\ \sqrt{|y+1|} - \frac{1}{2}y_1 - 1 \end{pmatrix}$$

the nonlinearity condition for $\bar{f}^{(2)}(\bar{y})$ at $\bar{y} = \bar{0}$ becomes (with $|y_1| \leq 1$)

$$\frac{|\bar{f}^{(2)}(\bar{y})|}{|\bar{y}|} = \frac{|\sqrt{|y_1+1|} - \frac{1}{2}y_1 - 1|}{|y_1| + |y_2|} \leq \frac{|\sqrt{|y_1+1|} - \frac{1}{2}y_1 - 1|}{|y_1|}$$
$$= \frac{|y_1|}{4\left|\sqrt{|y_1+1|} + \frac{y_1}{2} + 1\right|}$$

The latter is not greater than $|y_1|/2$ for $|y_1| \leq 1$. Thus since $d^2y/dt^2 + dy/dt + (\frac{1}{2})y = 0$ is asymptotically stable at $y = 0$, (14) is asymptotically stable at $x = 1$. It should be observed that the dominant part of Eq. (15) near $y = 0$ is characterized by the nonlinearity condition and does not consist simply of the linear terms of (15).

EXERCISES

1. Consider the linear system

$$\frac{dx}{dt} = a_{11}x + a_{12}y$$
$$\frac{dy}{dt} = a_{21}x + a_{22}y$$

where each of the coefficients a_{11}, a_{12}, a_{21}, and a_{22} is a constant.

Characterize the stability of the origin in terms of the coefficients.

2. Justify the statements made in Examples 1 and 2.

3. Using Theorem 1 and Exercise 1, discuss the stability of each of the singular solutions of the nonlinear system

$$\frac{dx}{dt} = x^2 - y$$

$$\frac{dy}{dt} = x - y$$

Find a single second-order equation which is equivalent to this system.

4. Let $\bar{f}(\bar{x})$ be continuously differentiable, and let $\bar{c}$ satisfy $\bar{f}(\bar{c}) = \bar{0}$. Show that $\bar{f}(\bar{x}) = \bar{f}_{\bar{x}}(\bar{c})\,(\bar{x} - \bar{c}) + \bar{f}^{(1)}(\bar{x})$, where $\bar{f}_{\bar{x}}(\bar{c})$ is the Jacobian matrix $(\partial f_i/\partial x_j)$ evaluated for $\bar{x} = \bar{c}$ and $\bar{f}^{(1)}(\bar{x})$ satisfies the nonlinearity condition (5). Illustrate for the right-hand member of the system in Exercise 3 and for each singular point of the system.

5. Show that the trivial solution of the nonlinear equation $d^2x/dt^2 - (dx/dt)^3 + x = 0$ is not stable. Use the technique introduced in Chap. 1 and show that the quantity $x^2 + (dx/dt)^2$ is nondecreasing along each trajectory and cannot be bounded. Explain the significance of such an example.

6. Show that the trivial solution of the nonlinear equation $d^2x/dt^2 + (dx/dt)^3 + x = 0$ is asymptotically stable. Explain the significance of such an example.

7. Show directly that the trivial solution of the Duffing equation $d^2x/dt^2 + \omega^2 x + \beta x^3 = 0$ is stable if $\omega^2 > 0$. Show that all other solutions are unstable if $\beta \neq 0$. Do any solutions other than the trivial one possess orbital stability?

3. *Stability of Singular Points of Nonautonomous Systems*

Using the representation theorem of Chap. 4 for linear periodic systems, one encounters no difficulty in generalizing the proof of Theorem 1 to a proof of Theorem 2.

THEOREM 2

If the periodic system

$$\frac{d\bar{x}}{dt} = P(t)\,\bar{x}$$

is asymptotically stable at the origin, then the system

$$\frac{d\bar{x}}{dt} = P(t)\,\bar{x} + \bar{f}^{(1)}(\bar{x}) \tag{16}$$

is asymptotically stable at the origin, provided $\bar{f}^{(1)}(\bar{x})$ satisfies the nonlinearity condition

$$\lim_{\bar{x}\to\bar{0}} \frac{|\bar{f}^{(1)}(\bar{x})|}{|\bar{x}|} = 0$$

EXAMPLE 6

Consider the equation

$$\frac{d^2x}{dt^2} + f(x)\frac{dx}{dt} + (\omega^2 + \epsilon \cos t)x = 0$$

Assume that $f(0) = k > 0$. We have the equivalent vector system

$$\frac{d\bar{x}}{dt} = P(t)\,\bar{x} + \bar{f}^{(1)}(\bar{x})$$

where

$$P(t) = \begin{pmatrix} 0 & 1 \\ -\omega^2 - \epsilon \cos t & -k \end{pmatrix} \quad \text{and} \quad \bar{f}^{(1)}(\bar{x}) = \begin{pmatrix} 0 \\ [k - f(x_1)]x_2 \end{pmatrix}$$

Now

$$\frac{|\bar{f}^{(1)}(\bar{x})|}{|\bar{x}|} = \frac{|k - f(x_1)|\,|x_2|}{|x_1| + |x_2|} \le |k - f(x_1)|$$

and so if $f(x)$ is continuous at $x = 0$, $\bar{f}^{(1)}(\bar{x})$ satisfies the non-linearity condition.

For small $|\epsilon|$, the linear equation

$$\frac{d^2x}{dt^2} + k\frac{dx}{dt} + (\omega^2 + \epsilon \cos t)x = 0$$

is asymptotically stable at the origin; hence, the nonlinear equation is asymptotically stable at the origin.

◆ ◆ ◆ ◆ ◆ ◆ ◆

There are many cases in which the autonomous form of the nonlinear term in either (4) or (16) is a serious restriction. It is clear from the proof of Theorem 1 that if the nonlinearity condition (5) is replaced by

$$\lim_{\bar{x} \to \bar{c}} \frac{|\bar{f}^{(1)}(\bar{x},t)|}{|\bar{x} - \bar{c}|} = 0 \tag{17}$$

uniformly for $t \geq 0$, then the conclusions of Theorems 1 and 2 still hold. However, this is often not enough, and in many cases the limit in (17) cannot be expected to be uniform for $t \geq 0$. If we strengthen the nonlinearity condition somewhat so that the ratio $|\bar{f}^{(1)}(\bar{x},t)| / |\bar{x} - \bar{c}|$ vanishes like some positive power of $|\bar{x} - \bar{c}|$, then it is possible to relax the uniform requirement in t and to permit a growth factor. In fact, we have the rather interesting result, Theorem 3.

THEOREM 3

Let $\bar{f}^{(1)}(\bar{x},t)$, for $|\bar{x}| \leq \delta$ with $\delta > 0$, satisfy the inequality

$$\frac{|\bar{f}^{(1)}(\bar{x},t)|}{|\bar{x}|} \leq m e^{\alpha t} |\bar{x}| \tag{18}$$

for certain positive constants m and α and all $t \geq 0$. Let $A(t)$ be either a constant or periodic matrix, and let the principal matrix solution $Y(t)$ of the linear system

$$\frac{d\bar{x}}{dt} = A(t) \bar{x} \tag{19}$$

satisfy $|Y(t)| \leq c_1 e^{-\alpha_1 t}$

for suitable positive constants c_1 and α_1. That is, let $A(t)$ be either a constant or periodic matrix for which the trivial solution of (19) is asymptotically stable.   Then if $\alpha < \alpha_1$, the nonlinear system

$$\frac{d\bar{x}}{dt} = A(t)\,\bar{x} + \bar{f}^{(1)}(\bar{x},t) \tag{20}$$

is asymptotically stable at the origin.

PROOF: We consider only the case where $A(t)$ is a constant matrix.   The proof for the periodic case is similar.   From (44) of Chap. 4, we have for any solution of (20),

$$\bar{x}(t) = Y(t)\,\bar{c} + \int_0^t Y(t-s)\,\bar{f}^{(1)}(\bar{x}(s),\,s)\,ds$$

where $\bar{c}$ is initial vector.   Hence

$$|\bar{x}(t)| \le c_1 |\bar{c}|\,e^{-\alpha_1 t} + c_1 \int_0^t e^{-\alpha_1(t-s)}\,|\bar{f}^{(1)}(\bar{x}(s),\,s)|\,ds$$

or, what is the same,

$$|\bar{x}(t)|\,e^{\alpha_1 t} \le c_1 |\bar{c}| + c_1 \int_0^t |\bar{f}^{(1)}(\bar{x}(s),\,s)|\,e^{\alpha_1 s}\,ds$$

So long as the solution $\bar{x}(t)$ satisfies $|\bar{x}(t)| \le \delta$, we have further, using (18),

$$[|\bar{x}(t)|\,e^{\alpha_1 t}] \le c_1 |\bar{c}| + c_1 m \int_0^t |\bar{x}(s)|\,e^{\alpha s}[|\bar{x}(s)|\,e^{\alpha_1 s}]\,ds \tag{21}$$

If we require that $|\bar{c}| < \delta|c_1$, so that (21) is valid over some positive interval of time, then using Gronwall's lemma, (21) yields

$$|\bar{x}(t)|\,e^{\alpha_1 t} \le c_1 |\bar{c}| \exp\left[c_1 m \int_0^t |\bar{x}(s)|\,e^{\alpha s}\,ds\right]$$

or, what is the same,

$$\Phi(t) \le |\bar{c}| \exp\left[c_1{}^2 m \int_0^t \Phi(s)\,ds - (\alpha_1 - \alpha)t\right] \tag{22}$$

where $\Phi(t) = |\bar{x}(t)|\,e^{\alpha t}/c_1$.   Let us further require that $|\bar{c}| < (\alpha_1 - \alpha)/mc_1$ so that $\Phi(t) < (\alpha_1 - \alpha)/mc_1{}^2$ on some interval $0 \le t \le b$, $b > 0$.   Then (22), in turn, yields

$$\frac{|\bar{x}(t)|}{c_1} \le \Phi(t) \le |\bar{c}|$$

since the exponent is initially negative.   Thus, the two restrictions $|\bar{c}| \leq \delta/c_1$ and $|\bar{c}| < (\alpha_1 - \alpha)/mc_1$ imply that the inequalities $|\bar{x}(t)| \leq \delta$ and $\Phi(t) \leq |\bar{c}|$ hold throughout and, because of the latter, that $\lim_{t \to \infty} |\bar{x}(t)| = 0$.

Example 7

Consider the nonlinear scalar equation $dx/dt + 2x - e^t x^2 = 0$. The linear equation $dx/dt + 2x = 0$ is asymptotically stable at the origin and $Y = e^{-2t}$ is its principal "matrix" solution.   Thus since $|e^t x^2|/|x| = |x|e^t$, the nonlinear system is also asymptotically stable at the origin.   Indeed, we can solve it explicitly.   For upon multiplying by the integrating factor e^{2t}, we obtain

$$\frac{d(xe^{2t})}{dt} = e^{-t}(e^{2t}x)^2$$

which is separable.   Thus for any nontrivial solution,

$$x = \frac{1}{(1/x_0 - 1)\, e^{2t} + e^t}$$

where x_0 is the initial value of x for $t = 0$.   Clearly, if $x_0 \leq +1$, $\lim_{t \to \infty} x = 0$ so that indeed the origin is asymptotically stable.   If $1 < x_0$, the corresponding solution exists only for a finite interval of time.

EXERCISES

1. Prove Theorem 2.

2. Prove Theorem 3 when the linear approximation is periodic.

3. State and prove a result similar to that in Theorem 3 for a system (20) satisfying the nonlinearity condition

$$\frac{|\bar{f}^{(1)}(\bar{x},t)|}{|\bar{x}|} \leq me^{\alpha t}\,|\bar{x}|^\beta$$

with m, α, and β certain positive constants.

4. Explain why the trivial solution of the nonlinear equation $d^2x/dt^2 + k\,dx/dt - t^N(dx/dt)^2 + x = 0$ is asymptotically stable for any positive number N provided the constant k is positive.

5. Explain why the trivial solution of the nonlinear equation $d^2x/dt^2 + k\,dx/dt + (x - t^Nx^3) = 0$ is asymptotically stable for any positive number N provided the constant k is positive.

6. Discuss the particular significance of Theorem 3 for systems which are analytic (or sufficiently differentiable) in $\bar{x}$.

7. Rephrase Theorem 1 for nonautonomous systems satisfying (17).

8. Using the result in Exercise 7, introduce a suitable artifice to show that the nonlinear system $d\bar{x}/dt = A(\bar{\beta})\,\bar{x} + \bar{f}^{(1)}(\bar{\beta},\bar{x},t)$, where $\bar{\beta}$ is a vector parameter, is asymptotically stable at the origin if the linear approximation $d\bar{x}/dt = A(\bar{\beta})\,\bar{x}$ is. Introduce suitable restrictions on the perturbational term $\bar{f}^{(1)}(\bar{\beta},\bar{x},t)$ and the coefficient matrix $A(\bar{\beta})$. The nonlinear system is then said to possess *structural* stability near the origin.

9. Under what circumstances does the nonlinear equation $d^2x/dt^2 + k\,dx/dt + \omega^2x = \beta f(x,dx/dt)$ possess structural stability near the origin?

• • • • • • •

We conclude this section by considering a theorem which essentially generalizes to nonlinear systems a result obtained already in Chap. 4 for linear systems. We consider a nonlinear system in which the nonlinear part is impulsively small relative to the linear part.

THEOREM 4

Suppose δ and c are positive constants such that

$$\int_0^\infty \frac{\left|\bar{f}^{(1)}(\bar{x}(s),\,s)\right|}{\left|\bar{x}(s)\right|}\,ds \le c$$

for every continuous functions $\bar{x}(s)$ satisfying $\left|\bar{x}(s)\right| \le \delta$. Then if the linear system

$$\frac{d\bar{x}}{dt} = A(t)\,\bar{x} \tag{23}$$

where $A(t)$ is either a constant or periodic matrix, is stable at the origin, then the nonlinear system

$$\frac{d\bar{x}}{dt} = A(t)\,\bar{x} + \bar{f}^{(1)}(\bar{x},t) \tag{24}$$

is stable at the origin.

PROOF: Again we consider only the constant matrix case. The hypotheses imply that the principal matrix solution $Y(t)$ of (23) is bounded as $t \to \infty$. Thus, using (44) of Chap. 4, we have, for any solution of (24)

$$|\,\bar{x}(t)\,| \le c_1\,|\,\bar{c}\,| + c_1 \int_0^t \frac{|\,\bar{f}^{(1)}(\bar{x}(s),\,s)\,|}{|\,\bar{x}(s)\,|}|\,\bar{x}(s)\,|\,ds \tag{25}$$

for suitable c_1, where $\bar{c}$ is the initial vector of $\bar{x}$. If $|\,\bar{c}\,| < \delta$, then $|\,\bar{x}(s)\,| < \delta$ will hold on some interval $0 \le s \le b,\ b > 0$. Using Gronwall's lemma, (25) yields

$$|\,\bar{x}(t)\,| \le c_1\,|\,\bar{c}\,| \exp\left[\int_0^t \frac{|\,\bar{f}^{(1)}(\bar{x}(s),\,s)\,|}{|\,\bar{x}(s)\,|}\,ds\right] \le c_1\,|\,\bar{c}\,|\,e^c$$

Thus, if $\bar{c}$ further satisfies $|\,\bar{c}\,| \le \delta/c_1 e^c$, then the inequality $\bar{x}(t)\,| \le \delta$ will hold for *all* $t \ge 0$ and the theorem is proved.

This theorem generalizes one of the linear cases considered in Chap. 4, namely, $\bar{f}^{(1)}(\bar{x},t) = B(t)\,\bar{x}$, with $B(t)$ impulsively small as $t \to \infty$.

EXAMPLE 8

Consider the nonlinear equation

$$\frac{d^2x}{dt^2} + \left[1 - \frac{f(x)}{1 + t^2}\right]x = 0,$$

where $f(x)$ is continuous. Since the linear approximation $d^2x/dt^2 + x = 0$ is stable at the origin and for any continuous functions $x_1(t)$ and $x_2(t)$ satisfying, say, $|x_1| + |x_2| \leq 1$ for all t, we have

$$\int_0^\infty \frac{|x_1(t)|\ |f(x_1(t))|}{|x_1(t)| + |x_2(t)|}\ \frac{dt}{1 + t^2} \leq \int_0^\infty \frac{|f(x_1(t))|}{1 + t^2}\ dt$$

$$\leq M \int_0^\infty \frac{dt}{1 + t^2} = \frac{\pi}{2}\ M$$

where $M = \max\limits_{-1 \leq x \leq +1} |f(x)|$, it follows from Theorem 4 that the nonlinear equation is stable at the origin.

4. Stability of Singular Solutions of Implicit Equations

The technique used in the preceding sections may be applied to an implicit equation of the form

$$\frac{d\bar{x}}{dt} = A\bar{x} + \bar{f}\left(\bar{x}, \frac{d\bar{x}}{dt}\right) \tag{26}$$

if $\bar{f}(\bar{x}, \bar{y})$ satisfies the condition

$$\lim_{\substack{\bar{x} \to \bar{0} \\ \bar{y} \to \bar{0}}} \frac{|\bar{f}(\bar{x}, \bar{y})|}{|\bar{x}| + |\bar{y}|} = 0 \tag{27}$$

We limit the discussion to autonomous systems merely for convenience of expression. The condition (27) implies that the derivative in (26) is "nearly explicit," at least near $\bar{x} = \bar{0}$ and $d\bar{x}/dt = 0$. To identify those solutions of (26) of which we shall speak, we assume that $\bar{x} = \bar{0}$, $\bar{y} = \bar{0}$ is a solution of the algebraic equation

$$\bar{y} = A\bar{x} + \bar{f}(\bar{x}, \bar{y}) \tag{28}$$

and that (28) defines $\bar{y}$ as a single-valued Lipschitzian function of $\bar{x}$ in a neighborhood of $\bar{x} = \bar{0}$. We have then, Theorem 5.

THEOREM 5

If the linear system

$$\frac{d\bar{x}}{dt} = A\bar{x} \tag{29}$$

is asymptotically stable at the origin, and if $\bar{f}(\bar{x},\bar{y})$ satisfies the nonlinearity condition (27), then (26) is asymptotically stable at the origin.

PROOF: Again using (44) of Chap. 4, we have for any solution of (26),

$$\bar{x}(t) = Y(t)\,\bar{c} + \int_0^t Y(t-s)\,\bar{f}(\bar{x}(s),\,\bar{y}(s))\,ds \qquad (30)$$

where $Y(t)$ is the principal matrix solution associated with (29), $\bar{y} = d\bar{x}/dt$, and $\bar{c}$ is an initial vector.   The hypotheses imply that

$$|Y(t)| \le c_1 e^{-\alpha t}$$

for suitable positive constants c_1 and α.   Hence (30) yields

$$|\bar{x}(t)| \le c_1|\bar{c}|\,e^{-\alpha t} + c_1 \int_0^t e^{-\alpha(t-s)} \frac{|\bar{f}(\bar{x}(s),\,\bar{y}(s))|}{|\bar{x}(s)|} |\bar{x}(s)|\,ds \quad (31)$$

On the other hand, from (26), we obtain

$$|\bar{y}| \le |A|\,|\bar{x}| + |\bar{f}(\bar{x},\bar{y})| \le c_2|\bar{x}| + |\bar{f}(\bar{x},\bar{y})|$$
$$= |\bar{x}|\left(c_2 + \frac{|\bar{f}(\bar{x},\bar{y})|}{|\bar{x}|}\right) \quad (32)$$

for a suitable positive constant c_2.   Thus

$$\frac{|\bar{y}|}{|\bar{x}|} \le c_2 + \frac{|\bar{f}(\bar{x},\bar{y})|}{|\bar{x}|}$$

and since

$$\frac{|\bar{f}(\bar{x},\bar{y})|}{|\bar{x}|} = \frac{|\bar{f}(\bar{x},\bar{y})|}{|\bar{x}|+|\bar{y}|}\left(1 + \frac{|\bar{y}|}{|\bar{x}|}\right)$$

it follows that

$$\frac{|\bar{f}(\bar{x},\bar{y})|}{|\bar{x}|} \le \frac{|\bar{f}(\bar{x},\bar{y})|}{|\bar{x}|+|\bar{y}|}\left[1 + c_2 + \frac{|\bar{f}(\bar{x},\bar{y})|}{|\bar{x}|}\right]$$

From this, in turn, we obtain the inequality

$$\frac{|\bar{f}(\bar{x},\bar{y})|}{|\bar{x}|} \le (1+c_2)\frac{|\bar{f}(\bar{x},\bar{y})|}{|\bar{x}|+|\bar{y}|}\left[\frac{1}{1 - |\bar{f}(\bar{x},\bar{y})|/(|\bar{x}|+|\bar{y}|)}\right] \quad (33)$$

provided $|\bar{f}(\bar{x},\bar{y})|/(|\bar{x}|+|\bar{y}|) < 1$.

Now let ϵ $(0 < \epsilon < 1)$ satisfy the inequality

$$c_1(1 + c_2) \frac{\epsilon}{1 - \epsilon} < \alpha \tag{34}$$

and let $\delta > 0$ be chosen, according to (27), so that

$$\frac{|\bar{f}(\bar{x},\bar{y})|}{|\bar{x}| + |\bar{y}|} < \epsilon \tag{35}$$

whenever $|\bar{x}| + |\bar{y}| < \delta$. If $|\bar{c}|$ is sufficiently small, then the inequality $|\bar{x}(t)| + |\bar{y}(t)| = |\bar{x}(t)| + |d\bar{x}/dt| < \delta$ will be satisfied on some interval $0 \leq t \leq b$, with $b > 0$. Then (31), (33), and (35) together yield

$$[|\bar{x}(t)| e^{\alpha t}] \leq c_1|\bar{c}| + c_1(1 + c_2) \frac{\epsilon}{1 - \epsilon} \int_0^t [|\bar{x}(s)| e^{\alpha s}] \, ds$$

By Gronwall's lemma we obtain further

$$|\bar{x}(t)| e^{\alpha t} \leq c_1|\bar{c}| \exp \left[c_1(1 + c_2) \frac{\epsilon}{1 - \epsilon} \right] t$$

or, what is the same,

$$|\bar{x}(t)| \leq c_1|\bar{c}| \exp \left[c_1(1 + c_2) \frac{\epsilon}{1 - \epsilon} - \alpha \right] t \tag{36}$$

On the other hand, from (32) and (35) we have

$$|\bar{y}(t)| \leq |\bar{x}(t)| (c_2 + \epsilon)$$

which, together with (36), yields

$$|\bar{x}(t)| + |\bar{y}(t)| \leq c_1|\bar{c}| (1 + c_2 + \epsilon)$$
$$\exp \left[c_1(1 + c_2) \frac{\epsilon}{1 - \epsilon} - \alpha \right] t \tag{37}$$

According to (34), the factor $\left[c_1(1 + c_2) \frac{\epsilon}{1 - \epsilon} - \alpha \right]$ in the exponent of (37) is negative. Thus if $\bar{c}$ satisfies the inequality

$$|\bar{c}| < \frac{\delta}{c_1(1 + c_2 + \epsilon)}$$

(37) will remain valid for all $t \geq 0$ and certainly

$$\lim_{t \to \infty} \left[\, |\bar{x}(t)| + |\bar{y}(t)| \, \right] = 0$$

EXAMPLE 9

Consider the equation

$$(1 + x) \frac{d^2x}{dt^2} + \frac{dx}{dt} + x = 0$$

The linear approximation $d^2x/dt^2 + dx/dt + x = 0$ is asymptotically stable at the origin, and the nonlinear portion satisfies

$$\frac{|x| \, |d^2x/dt^2|}{|x| + 2|dx/dt| + |d^2x/dt^2|} \leq |x|$$

Thus according to Theorem 5, the nonlinear equation is also asymptotically stable at the origin.   One obtains the same result upon division by $1 + x$ using Theorem 1.   In fact from

$$\frac{d^2x}{dt^2} + \left(\frac{1}{1 + x} \right) \frac{dx}{dt} + \left(\frac{1}{1 + x} \right) x = 0$$

we are led to consider

$$\frac{d^2x}{dt^2} + \left(1 - \frac{x}{1 + x} \right) \frac{dx}{dt} + \left(1 - \frac{x}{1 + x} \right) x = 0$$

The nonlinear portion satisfies

$$\frac{(|x|/|1 + x|)(|dx/dt| + |x|)}{|x| + |dx/dt|} = \frac{|x|}{|1 + x|} < 2|x| \quad \text{for } |x| < \tfrac{1}{2}$$

which is the appropriate nonlinearity condition for this case.

EXAMPLE 10

Consider the equation

$$\sin \left(\frac{d^2x}{dt^2} \right) + \frac{dx}{dt} + x = 0 \tag{38}$$

Real solutions exist only for sufficiently small values of $|dx/dt +$

x| since $|\sin (d^2x/dt^2)| \leq 1$. For convenience let us restrict d^2x/dt^2 to the range $-\pi/2 < d^2x/dt^2 \leq \pi/2$. Then (38) possesses a unique solution through each initial point near the origin and is equivalent to the explicit equation

$$\frac{d^2x}{dt^2} = - \arcsin\left(x + \frac{dx}{dt} \right) \qquad (39)$$

where arcsin is the principal branch of the inverse sine function. Equation (38) may also be written in the form

$$\frac{d^2x}{dt^2} + \frac{dx}{dt} + x + \left[\sin\left(\frac{d^2x}{dt^2}\right) - \frac{d^2x}{dt^2} \right] = 0$$

Now

$$\frac{|\sin (d^2x/dt^2) - d^2x/dt^2|}{|x| + 2|dx/dt| + |d^2x/dt^2|} \leq \frac{|\sin (d^2x/dt^2) - d^2x/dt^2|}{|d^2x/dt^2|}$$

$$= \left| \frac{\sin (d^2x/dt^2)}{d^2x/dt^2} - 1 \right|$$

and from the well-known limit of $\sin \theta/\theta$ for $\theta \to 0$, we conclude that the nonlinear portion of (38) near the origin satisfies the nonlinearity condition. Thus, since the linear approximation $d^2x/dt^2 + dx/dt + x = 0$ is asymptotically stable at the origin, the same is true of (38).

EXERCISES

1. Prove Theorem 4 when the linear approximation is a periodic system.

2. Discuss, in detail, the application of Theorem 5 made in Examples 9 and 10. In particular, explain the origin of the nonlinearity condition in each case.

3. Using Theorem 1, show that the trivial solution of (39) is asymptotically stable.

4. Replace the nonlinearity condition (27) with the strong inequality

$$\frac{|\tilde{f}(\bar{x},\bar{y})|}{|\bar{x}| + |\bar{y}|} \leq m(|\bar{x}| + |\bar{y}|)$$

where m is a positive constant, and use this to prove Theorem 5.

5. Discuss the particular significance of the result in Exercise 4 for analytic (or sufficiently differentiable) systems.

5. *Stability of Nonsingular Solutions*

The stability characteristics of a nonsingular trajectory $\bar{x}(t)$ of the equation

$$\frac{d\bar{x}}{dt} = \bar{f}(\bar{x},t) \tag{40}$$

are reflected in the stability characteristics of a singular solution of a related equation. In fact, if $\bar{x}^*(t)$ is a neighboring solution, then the perturbation function $\bar{y}(t) = \bar{x}(t) - \bar{x}^*(t)$ satisfies the equation

$$\frac{d\bar{y}}{dt} = \bar{f}(\bar{x}(t), t) - \bar{f}(\bar{x}(t) - \bar{y}(t), t) = \bar{f}^*(\bar{y}(t), t) \tag{41}$$

where the right-hand member vanishes identically for $\bar{y}(t) = \bar{0}$. Thus $\bar{x}(t)$ is a stable solution of (40) if and only if the trivial solution of (41) is stable. Equation (41) is called the *perturbation equation*. Typically, one attempts to infer the stability characteristics not from the perturbation equation but from the *equation of first variation*. The latter is the unique linear part of (41) near the origin, assuming that $\bar{f}(\bar{x},t)$ is continuously differentiable in $\bar{x}$. We denote by $\bar{f}_{\bar{x}}(\bar{x},t)$ the Jacobian matrix

$$\frac{d\bar{f}}{d\bar{x}} = \begin{pmatrix} \dfrac{\partial f_1}{\partial x_1} & \dfrac{\partial f_1}{\partial x_2} & \cdots & \dfrac{\partial f_1}{\partial x_n} \\ \dfrac{\partial f_2}{\partial x_1} & \dfrac{\partial f_2}{\partial x_2} & \cdots & \dfrac{\partial f_2}{\partial x_n} \\ \cdots\cdots\cdots\cdots\cdots\cdots \\ \dfrac{\partial f_n}{\partial x_1} & \dfrac{\partial f_n}{\partial x_2} & \cdots & \dfrac{\partial f_n}{\partial x_n} \end{pmatrix} \tag{42}$$

where $f_1, f_2, \ldots, f_n$ are the components of $\bar{f}(\bar{x},t)$, and the linear part of (41) is

$$\frac{d\bar{y}}{dt} = \bar{f}_{\bar{x}}(\bar{x}(t),\ t)\ \bar{y}(t) \tag{43}$$

The perturbation equation (41) may then be expressed in the form

$$\frac{d\bar{y}}{dt} = \bar{f}_{\bar{x}}(\bar{x}(t),\ t)\ \bar{y}(t) + \bar{f}^{(1)}(\bar{y}(t),\ t) \tag{44}$$

and the theorem of the mean guarantees that $\bar{f}^{(1)}(\bar{y},t)$ satisfies the nonlinearity condition. Of course, the stability theory for even the linear (first variation) equation (43) is manifestly incomplete except when $\bar{f}_{\bar{x}}(\bar{x}(t),\ t)$ is either constant or periodic, or approximately constant or periodic. Still, these cases arise sufficiently often to justify the emphasis placed upon them in Chap. 4.

An important case arises when $\bar{x}(t)$ is a periodic solution, for when it is stable, it represents steady-state phenomena. For example, a periodic solution $\bar{x}(t)$ may be the steady-state response to a periodic input in (40). In this case, the equation of first variation (43) becomes

$$\frac{d\bar{y}}{dt} = P(t)\ \bar{y} \tag{45}$$

where $P(t) = \bar{f}_{\bar{x}}(\bar{x}(t),\ t)$ is a periodic matrix. If the origin of (45) is asymptotically stable, then the steady-state response $\bar{x}(t)$ is asymptotically stable.

Another case, frequently encountered, occurs when (40) is autonomous, i.e.,

$$\frac{d\bar{x}}{dt} = \bar{f}(\bar{x}) \tag{46}$$

and $\bar{x}(t)$ is a periodic (self-sustained) solution of (46). It may be that (46) is conservative and exhibits only periodic solutions. In any event, with $\bar{x}(t)$ periodic, the equation of first variation (43) is again periodic with the period of $\bar{x}(t)$. In this case, however, the equation of first variation is *never* asymptotically stable at the origin. For upon differentiating (46) with respect to t, we obtain

$$\frac{d(d\bar{x}/dt)}{dt} = \bar{f}_{\bar{x}}(\bar{x}) \frac{d\bar{x}}{dt}$$

and this shows that $d\bar{x}/dt$ is a solution of the equation of first variation. In particular, since $\bar{x}(t)$ is periodic, $d\bar{x}/dt$ is periodic with the period of $\bar{x}(t)$, and so the equation of first variation possesses a periodic solution with the period of the system. Therefore, the equation of first variation can be at best stable at the origin. It can be shown,[1] however, that if the equation of first variation possesses just one characteristic exponent with zero real part (corresponding to the known periodic solution $d\bar{x}/dt$) and each of the remaining characteristic numbers has a negative real part, then $\bar{x}(t)$ possesses asymptotic *orbital* stability. In fact, each solution $\bar{x}^*(t)$ near the periodic one possesses an asymptotic phase; i.e., there is a γ such that

$$\lim_{t \to \infty} |\bar{x}(t) - \bar{x}^*(t + \gamma)| = 0$$

EXAMPLE 11

Consider the forced Duffing equation

$$\frac{d^2x}{dt^2} + k\frac{dx}{dt} + x + \beta x^3 = \beta f_0 \cos \omega t \qquad (47)$$

where each of k, β, f_0, and ω is a constant. The vector equivalent of (47) is

$$\frac{d\bar{x}}{dt} = \bar{f}(\bar{x},t)$$

with $\qquad \bar{f}(\bar{x},t) = \begin{pmatrix} x_2 \\ -x_1 - \beta x_1{}^3 - kx_2 + \beta f_0 \cos \omega t \end{pmatrix}$

---

[1] E. A. Coddington and N. Levinson, "Theory of Ordinary Differential Equations," pp. 323–327, McGraw-Hill Book Company, Inc., New York, 1955.

The equation of first variation is

$$\frac{d\bar{y}}{dt} = \tilde{f}_{\bar{x}}(\bar{x}(t), t)\bar{y} \tag{48}$$

where
$$\tilde{f}_{\bar{x}} = \begin{pmatrix} 0 & 1 \\ -1 - 3\beta x_1^2 & -k \end{pmatrix}$$

The scalar equivalent of (48) is

$$\frac{d^2y}{dt^2} + k\frac{dy}{dt} + [1 + 3\beta x^2(t)]y = 0 \tag{49}$$

where $x(t)$ is a scalar solution of (47). In particular, if $x(t)$ is a periodic solution of (47) and if $k > 0$, then Theorem 6 of Chap. 4 guarantees that the trivial solution of (49) is asymptotically stable, provided $|\beta|$ is sufficiently small. Because the equation of first variation is asymptotically stable at the origin, the corresponding periodic solution of (47) is asymptotically stable for $|\beta|$ small.

EXAMPLE 12.   POINCARÉ CRITERION FOR ORBITAL STABILITY

Consider the two-dimensional autonomous system

$$\begin{aligned} \frac{dx_1}{dt} &= p(x_1,x_2) \\ \frac{dx_2}{dt} &= q(x_1,x_2) \end{aligned} \tag{50}$$

where $p(x_1,x_2)$ and $q(x_1,x_2)$ are continuously differentiable in each variable. Suppose that this system possesses a nontrivial periodic solution $[x_1(t), x_2(t)]$ of period $\tau > 0$. The equations of first variation are

$$\begin{aligned} \frac{dy_1}{dt} &= p_1(x_1(t), x_2(t))y_1 + p_2(x_1(t), x_2(t))y_2 \\ \frac{dy_2}{dt} &= q_1(x_1(t), x_2(t))y_1 + q_2(x_1(t), x_2(t))y_2 \end{aligned} \tag{51}$$

where we use the notation $p_1 = \partial p/\partial x_1$, $p_2 = \partial p/\partial x_2$, $q_1 = \partial q/\partial x_1$, and $q_2 = \partial q/\partial x_2$. This is a linear periodic system of period τ, and therefore its principal matrix solution may be expressed in the form

$$Y(t) = Q(t) e^{Bt} \tag{52}$$

where $Q(t)$ is a 2-by-2 periodic matrix of period τ, and B is a 2-by-2 constant matrix. The determinant of $Y(t)$ satisfies

$$|Y(t)| = |Q(t)| \, |e^{Bt}| = \exp\left[\int_0^t \text{trace } P(s) \, ds \right] \tag{53}$$

where $P(t)$ is the 2-by-2 matrix of coefficients in (51). On the other hand, we have for the determinant of e^{Bt}

$$|e^{Bt}| = \exp(\text{trace } Bt) \tag{54}$$

From (52) it is clear that $Q(0) = I$ (the identity matrix), and hence $Q(\tau) = I$. Thus (53) and (54) together yield, for $t = \tau$

$$\text{trace } B = \frac{1}{\tau} \int_0^\tau \text{trace } P(s) \, ds$$

$$= \frac{1}{\tau} \int_0^\tau [p_1(x_1(s), x_2(s)) + q_2(x_1(s), x_2(s))] \, ds \tag{55}$$

In general, the trace of a constant matrix is equal to the sum of its characteristic numbers, and so (55) may be expressed in the form

$$\lambda_1 + \lambda_2 = \frac{1}{\tau} \int_0^\tau [p_1(x_1(s), x_2(s)) + q_2(x_1(s), x_2(s))] \, ds \tag{56}$$

where λ_1 and λ_2 are the characteristic numbers of B. Now the periodic system (51) is known to possess a periodic solution of period τ, namely, $[dx_1/dt, dx_2/dt]$, and this is possible only if zero is a characteristic number of B. Thus, (56) may be further expressed as

$$\lambda = \frac{1}{\tau} \int_0^\tau [p_1(x_1(s), x_2(s)) + q_2(x_1(s), x_2(s))] \, ds \tag{57}$$

where λ is the one possibly nonzero characteristic number. In some cases both characteristic numbers are zero. However, if

$\lambda < 0$, then the periodic solution of (51) is asymptotically stable. In fact, there exists a base for solutions of the linear system (51) consisting of the periodic solution $(dx_1/dt, dx_2/dt)$ and a second solution of the form $[y_1(t) = a(t) e^{\lambda t}, y_2(t) = b(t) e^{\lambda t}]$ with each of $a(t)$ and $b(t)$ periodic of period τ. Thus, if $\lambda < 0$, every solution tends exponentially to a multiple of the periodic one as $t \to \infty$. Clearly, the variational system is stable at the origin but is not asymptotically stable at the origin. Thus, the stability of the (original) periodic solution $[x_1(t), x_2(t)]$ of the nonlinear system (50) remains in question.

To investigate this question further, let us depict perturbations in (50) in terms of perturbations in initial values. Specifically, let the initial values of the periodic solution $[x_1(t), x_2(t)]$ be c_1 and c_2, and let the initial values of a neighboring solution be $c_1 + \delta_1$ and $c_2 + \delta_2$. (It is not difficult to show that as the pair (δ_1, δ_2) varies in some neighborhood of $(0,0)$, all neighboring solutions sufficiently close to the periodic one are represented.) The neighboring solution is then given *approximately* by $[x_1(t) + y_1(t), x_2(t) + y_2(t)]$, where $[y_1(t), y_2(t)]$ is the solution of the variational equations (51) satisfying the initial conditions $y_1(0) = \delta_1$ and $y_2(0) = \delta_2$. If these initial conditions can be met with the exponentially decaying portion of the general solution of (51) alone, then it would appear that the approximate solution $[x_1(t) + y_1(t), x_2(t) + y_2(t)]$ of (50) tends to the periodic one exponentially. However, *two* initial conditions cannot in general be met with a one-parameter solution. Therefore, we introduce another degree of freedom by way of the following artifice.

For any constant γ, the pair $[x_1(t + \gamma), x_2(t + \gamma)]$ is once again the periodic solution of (50), parameterized in a somewhat different fashion. Hence, a neighboring solution is given approximately by $[x_1(t + \gamma) + y_1(t), x_2(t + \gamma) + y_2(t)]$ as before, except that now the neighboring initial values become $x_1(\gamma) + y_1(0)$ and $x_2(\gamma) + y_2(0)$ in place of $c_1 + y_1(0)$ and $c_2 + y_2(0)$. We duplicate the given initial values $c_1 + \delta_1$ and $c_2 + \delta_2$ for the neighboring solution by requiring $y_1(0) = c_1 + \delta_1 - x_1(\gamma) = \delta_1^*$ and $y_2(0) = c_2 + \delta_2 - x_2(\gamma) = \delta_2^*$. For $\gamma = 0$, these are as before, but now we may vary γ which, in turn, induces variations in the initial values δ_1^* and δ_2^* of $y_1(t)$ and $y_2(t)$, respectively. To be sure, the variations are not arbitrary since $x_1(\gamma)$ and $x_2(\gamma)$ are given functions. Nonetheless, the variations are sufficient to ensure that only the exponentially decaying portion of the general solution of (51) is required.

To see this, we note that permissible initial values δ_1^* and δ_2^* form a one-parameter family $\delta_1^* b(0) = \delta_2^* a(0)$, since they must be in proportion to the initial values of the exponentially decaying solution $[y_1(t) = a(t)\ e^{\lambda t},\ y_2(t) = b(t)\ e^{\lambda t}]$. Hence we must be able to choose δ_1^* (say) and γ so that

$$x_1(\gamma) + \delta_1^* = \delta_1 + c_1$$
$$x_2(\gamma) + \frac{b(0)}{a(0)}\, \delta_1^* = \delta_2 + c_2 \tag{58}$$

with each of c_1, c_2, δ_1, and δ_2 given. The circumstance $a(0) = 0$ is not covered here but requires only an obvious modification. Clearly, for $\gamma = 0$ and $\delta_1^* = 0$, (58) yields $\delta_1 = 0$ and $\delta_2 = 0$. Hence, if the Jacobian

$$J = \begin{vmatrix} \dfrac{dx_1}{d\gamma} & 1 \\[2ex] \dfrac{dx_2}{d\gamma} & \dfrac{b(0)}{a(0)} \end{vmatrix} = \frac{b(0)}{a(0)} \frac{dx_1}{d\gamma} - \frac{dx_2}{d\gamma} \tag{59}$$

does not vanish for $\gamma = 0$, (58) will possess a (unique) solution (γ, δ_1^*) for each pair (δ_1, δ_2) with $|\delta_1|$ and $|\delta_2|$ sufficiently small. In the event $J = 0$ at the initial point, we merely move to another for which the right-hand member in (59) is *not* zero. That there exists such a point is clear since the ratio

$$\frac{dx_2/d\gamma}{dx_1/d\gamma} = \frac{dx_2}{dx_1}$$

which is the slope of the tangent to the trajectory, cannot be constant along a periodic solution. Finally, we note that the illusive γ sought here is an asymptotic phase for the neighboring solution.

EXAMPLE 13

Consider the Rayleigh equation

$$\frac{d^2x}{dt^2} - \mu\left[\frac{dx}{dt} - \frac{1}{3}\left(\frac{dx}{dt}\right)^3\right] + x = 0 \tag{60}$$

It is well known that for $\mu > 0$, (60) possesses a unique periodic solution which for μ small is given approximately by $x(t) = 2\cos t$.

The equations of first variation are

$$\begin{aligned}
\frac{dy_1}{dt} &= y_2 \\
\frac{dy_2}{dt} &= -y_1 + \mu\left[1 - \left(\frac{dx}{dt}\right)^2\right]y_2
\end{aligned} \tag{61}$$

Using (57), we have

$$\lambda = \frac{\mu}{\tau}\int_0^\tau\left[1 - \left(\frac{dx}{dt}\right)^2\right]dt$$

or, approximately,

$$\lambda = \frac{\mu}{2\pi}\int_0^{2\pi}(1 - 4\sin^2 t)\,dt = -\mu$$

for the one nonzero characteristic exponent of (61). Thus, it appears that the periodic solution of (60) possesses asymptotic orbital stability, at least for small μ. Actually, it does so for all $\mu > 0$.

EXERCISES

1. Explain why it is that the nonlinear term in (44) satisfies the nonlinearity condition.

2. Construct a variety of examples to illustrate the concepts of perturbation equation, equation of first variation, and orbital stability as discussed in this section.

3. Discuss Example 12, supplying complete details for each of the several conclusions.   Note in particular those conclusions which might be of general interest, and illustrate each.   Supply figures to illustrate those steps which are essentially of a geometric nature.

4. Explain why there exists a nonsingular matrix T such that

$$B^* = TBT^{-1} = \begin{pmatrix} 0 & 0 \\ 0 & \lambda \end{pmatrix}$$

where B is the matrix in the exponent in (52).

5. Using the nonsingular matrix T of Exercise 4, introduce a linear transformation in (51), and show that the corresponding principal matrix solution $Y^*(t)$ in the new coordinates is of the form

$$Y^*(t) = Q^*(t) \begin{pmatrix} 1 & 0 \\ 0 & e^{\lambda t} \end{pmatrix}$$

where $Q^*(t)$ is a 2-by-2 periodic matrix of period τ.   Explain why "permissible" initial values for the linear variations in the new coordinates correspond to the circumstance $\delta_1^* = 0$, $\delta_2^* \neq 0$.

6. Introduce the linear transformation of Exercise 5 throughout in Example 12, and rederive the main results.   Note the relatively simple forms for (58) and (59).   What is the geometrical meaning of the condition $J \neq 0$ in the new coordinates?

7. Using the results (and the notation) introduced in the above exercises, prove that the periodic solution of (50) is asymptotically orbitally stable if the characteristic number λ given in (57) is negative.   In fact, explain why, in the new coordinates, the *exact* perturbation vector $\bar{y}$ satisfies an integral equation of the form

$$\bar{y}(t) = Q^*(t) \begin{pmatrix} 0 \\ \delta_2^* e^{\lambda t} \end{pmatrix} + \int_0^t Q^*(t) \begin{pmatrix} 1 & 0 \\ 0 & e^{\lambda(t-s)} \end{pmatrix} Q^{*-1}(s) \bar{J}(\bar{y}(s))\, ds$$

(i)

where $\bar{f}(\bar{y})$ satisfies the nonlinearity condition

$$\lim_{|\bar{y}|\to 0} \frac{|\bar{f}(\bar{y})|}{|\bar{y}|} = 0 \tag{ii}$$

Note that the first term on the right in (i) is the linear perturbation $\bar{y}^*$, i.e., the corresponding solution of the equation of first variation, and that the integral equation (i) may be expressed in the form

$$\bar{y}(t) = \bar{y}^*(t) + \int_0^t Q^*(t) \begin{pmatrix} 0 & 0 \\ 0 & e^{\lambda(t-s)} \end{pmatrix} Q^{*-1}(s)\, \bar{f}(\bar{y}(s))\, ds$$

$$+ \int_0^t Q^*(t) \begin{pmatrix} 1 & 0 \\ 0 & 0 \end{pmatrix} Q^{*-1}(s)\, \bar{f}(\bar{y}(s))\, ds \tag{iii}$$

It will be found that the solution $\bar{y}(t)$ satisfies an inequality of the form $|\bar{y}(t)| \le |\delta_2^*| K e^{\lambda t/2}$. Anticipating this, show that the second term on the right in (iii) necessarily tends to zero (the zero vector) as $t \to \infty$, provided $|\delta_2^*|$ is sufficiently small. First explain why $|\bar{f}(\bar{y}(s))| \le |\delta_2^*| K e^{\lambda s/2}$, provided $|\delta_2^*|$ is sufficiently small, and then proceed with the second term on the right in (iii). Now explain why the third term on the right in (iii) must also tend to zero (the zero vector) as $t \to \infty$ and, therefore, why (iii) may be reformulated (for present purposes) in the form

$$\bar{y}(t) = \bar{y}^*(t) + \int_0^t Q^*(t) \begin{pmatrix} 0 & 0 \\ 0 & e^{\lambda(t-s)} \end{pmatrix} Q^{*-1}(s)\, \bar{f}(\bar{y}(s))\, ds$$

$$- \int_t^\infty Q^*(t) \begin{pmatrix} 1 & 0 \\ 0 & 0 \end{pmatrix} Q^{*-1}(s)\, \bar{f}(\bar{y}(s))\, ds \tag{iv}$$

A solution of (iv) may be obtained by the method of successive approximations provided the initial vector is sufficiently small. Let $\bar{y}^{(1)}(t) = \bar{y}^*(t)$ be the first estimate and

$$\bar{y}^2(t) = \bar{y}^*(t) + \int_0^t Q^*(t) \begin{pmatrix} 0 & 0 \\ 0 & e^{\lambda(t-s)} \end{pmatrix} Q^{*-1}(s)\, \bar{f}(\bar{y}^{(1)}(s))\, ds$$

$$- \int_t^\infty Q^*(t) \begin{pmatrix} 1 & 0 \\ 0 & 0 \end{pmatrix} Q^{*-1}(s)\, \bar{f}(\bar{y}^{(1)}(s))\, ds \tag{v}$$

be the second.   Note that

$$\left|\, \bar{y}^{(1)}(t)\,\right| = c|\delta_2^*|e^{\lambda t} \le c|\delta_2^*|e^{\lambda t/2} \le c|\delta_2^*|$$

for a suitable positive constant c.   Let ϵ be a positive number and explain why there is a positive number δ such that

$$\left|\, \bar{f}(\bar{y})\,\right| \le \epsilon\left|\,\bar{y}\,\right| \qquad \text{for } \left|\,\bar{y}\,\right| \le \delta$$

Use this to derive from (v) the inequalities

$$\left|\, \bar{y}^{(2)}(t) - \bar{y}^{(1)}(t)\,\right| \le \epsilon|\delta_2^*|c_1c\left[\int_0^t e^{\lambda(t-s)}e^{\lambda s/2}\,ds + \int_t^\infty e^{\lambda s/2}\,ds\right]$$

$$\le \frac{4\epsilon|\delta_2^*|c_1c}{|\lambda|}\,e^{\lambda t/2}$$

and $$\left|\, \bar{y}^{(2)}(t)\,\right| \le c|\delta_2^*|\left(1 + \frac{4\epsilon c_1}{|\lambda|}\right)e^{\lambda t/2}$$

for a suitable positive constant c_1 and for $|\delta_2^*| < \delta/c$.   Then show that if

$$\bar{y}^{(3)}(t) = \bar{y}^*(t) + \int_0^t Q^*(t)\begin{pmatrix} 0 & 0 \\ 0 & e^{\lambda(t-s)} \end{pmatrix} Q^{*-1}(s)\,\bar{f}(\bar{y}^{(2)}(s))\,ds$$

$$- \int_t^\infty Q^*(t)\begin{pmatrix} 1 & 0 \\ 0 & 0 \end{pmatrix} Q^{*-1}(s)\,\bar{f}(\bar{y}^{(2)}(s))\,ds$$

is the third estimate, one may derive the inequalities

$$\left|\, \bar{y}^{(3)}(t) - \bar{y}^{(1)}(t)\,\right| \le \frac{4\epsilon|\delta_2^*|c_1c}{|\lambda|}\left(1 + \frac{4\epsilon c_1}{|\lambda|}\right)e^{\lambda t/2}$$

and $$\left|\, \bar{y}^{(3)}(t)\,\right| \le c|\delta_2^*|\left[1 + \frac{4\epsilon c_1}{|\lambda|}\left(1 + \frac{4\epsilon c_1}{|\lambda|}\right)\right]e^{\lambda t/2}$$

Derive analogous inequalities for each of the successive approximations and show that for ϵ sufficiently small and for a suitable positive constant K, $\left|\, \bar{y}^{(k)}(t)\,\right| \le |\delta_2^*|Ke^{\lambda t/2}$, for all $k = 1, 2, \ldots$. Finally, let ρ be a positive number and explain why $\bar{f}(\bar{y})$ satisfies the Lipschitz condition $\left|\, \bar{f}(\bar{y}) - \bar{f}(\bar{x})\,\right| \le \rho\left|\,\bar{y} - \bar{x}\,\right|$, in a sufficiently small neighborhood of the origin, and show that for $|\delta_2^*|$

sufficiently small, the successive approximations satisfy the inequalities

$$\left| \bar{y}^{(k+1)}(t) - \bar{y}^{(k)}(t) \right| \le \rho c_1 \int_0^t \left| \bar{y}^{(k)}(s) - \bar{y}^{(k-1)}(s) \right| ds$$
$$+ \rho c_1 \int_t^\infty \left| \bar{y}^{(k)}(s) - \bar{y}^{(k-1)}(s) \right| ds = \rho c_1 \int_0^\infty \left| \bar{y}^{(k)}(s) - \bar{y}^{(k-1)}(s) \right| ds$$

for $k = 2, 3, \ldots$ Now complete the argument.

6. *Direct Stability Method*

The methods of the preceding sections typically fail to indicate stability in the absence of asymptotic stability.   This is because the theory is developed around *comparison theorems*, the application of which requires that the linear part of a system be responsible for the stability.   Often, this is simply not the case.

A second method of Liapunov treats the stability problem directly and in a natural, though somewhat nonconstructive, way.   Application of the second method is not limited by an artificial circumstance as is the first (comparison) method. Rather, the second method is limited by practical matters relating to the construction or discovery of suitable testing functions.   Generally speaking, for the direct method one seeks something resembling a potential function which has a minimum at a given singular (equilibrium) point, say, the origin.   The idea is more or less that if one is able to find a suitable continuous scalar function $V(\bar{x})$ which is positive for $\bar{x} \ne 0$ and zero for $\bar{x} = 0$, then along a trajectory $\bar{x}(t)$ an inequality of the form $V(\bar{x}(t)) \le \delta$ will guarantee a related inequality $\left| \bar{x}(t) \right| \le \epsilon$ with $\epsilon \to 0$ as $\delta \to 0$.   The quantity $V(\bar{x})$ is then a measure of the smallness of $\bar{x}$.   More precisely, we define the scalar function $V(\bar{x})$ to be a *Liapunov function* for the system

$$\frac{d\bar{x}}{dt} = \bar{f}(\bar{x}) \tag{62}$$

if for $\bar{x}$ in some neighborhood of the origin, $V(\bar{x})$ is positive for $\bar{x} \neq 0$, zero for $\bar{x} = 0$, has continuous first derivatives in each variable, and satisfies

$$\nabla V(\bar{x})\, \bar{f}(\bar{x}) \leq 0 \tag{63}$$

where $\nabla V(\bar{x})$ is the gradient of $V(\bar{x})$ expressed as a row vector $\left(\dfrac{\partial V}{\partial x_1}, \dfrac{\partial V}{\partial x_2}, \cdots, \dfrac{\partial V}{\partial x_n}\right)$. This last inequality guarantees that the quantity $V(\bar{x}(t))$ is nonincreasing along each trajectory $\bar{x}(t)$ sufficiently near the origin. In fact, the total t derivative is precisely

$$\frac{d}{dt}\,[V(\bar{x}(t))] \,=\, \nabla V(\bar{x}(t))\, \frac{d\bar{x}}{dt} \,=\, \nabla V(\bar{x}(t))\, \bar{f}(\bar{x}(t))$$

and, therefore, nonpositive. We have the elementary but fundamental Theorem 6.

THEOREM 6

If there exists a Liapunov function $V(\bar{x})$ for (62), then (62) is stable at the origin.

PROOF: Let ϵ be a positive number such that (63) and the other hypotheses concerning $V(\bar{x})$ hold for $|\bar{x}| \leq \epsilon$. Since $V(\bar{x}) > 0$ on the surface $|\bar{x}| = \epsilon$ and the surface is compact, i.e., closed and bounded, $V(\bar{x})$ is bounded away from zero on $|\bar{x}| = \epsilon$. Suppose $V(\bar{x}) \geq \eta > 0$ for $|\bar{x}| = \epsilon$. Then there exists a $\delta > 0$ such that, say $V(\bar{x}) \leq \eta/2$ for $|\bar{x}| \leq \delta$. Surely $\delta < \epsilon$. If a trajectory $\bar{x}(t)$ of (62) satisfies $\bar{x}(0) = \bar{c}$ with $|\bar{c}| \leq \delta$, then since V is nonincreasing along the trajectory $\bar{x}(t)$, we have

$$V(\bar{x}(t)) \leq V(\bar{c}) \leq \frac{\eta}{2}$$

so long as $|\bar{x}(t)| \leq \epsilon$. But $\bar{x}(t)$ cannot then penetrate the surface $|\bar{x}| = \epsilon$ on which we have $V(\bar{x}) \geq \eta$. Hence, $|\bar{x}(t)| \leq \epsilon$ for all $t \geq 0$, which proves the theorem.

EXAMPLE 14

Consider the second-order equation

$$\frac{d^2x}{dt^2} + \frac{k\,dx}{dt} + \omega^2 x = 0$$

and the related vector quantities

$$\bar{x} = \begin{pmatrix} x_1 \\ x_2 \end{pmatrix} = \begin{pmatrix} x \\ \dfrac{dx}{dt} \end{pmatrix} \quad \text{and} \quad \bar{f}(\bar{x}) = \begin{pmatrix} x_2 \\ -\omega^2 x_1 - kx_2 \end{pmatrix}$$

Let $V(\bar{x}) = \omega^2 x_1^2 + x_2^2$ for which, certainly, $V(\bar{x}) > 0$ for $\bar{x} \neq \bar{0}$ and $V(\bar{0}) = 0$. Also from the above we have

$$\nabla V(\bar{x})\,\bar{f}(\bar{x}) = (2\omega^2 x_1,\, 2x_2) \begin{pmatrix} x_2 \\ -\omega^2 x_1 - kx_2 \end{pmatrix} = -2kx_2^2$$

Thus if $k \geq 0$, $V(\bar{x})$ is a Liapunov function and the system is stable at the origin. More generally, if $f(x,y) \geq 0$ near the origin, then the nonlinear equation

$$\frac{d^2x}{dt^2} + f\left(x, \frac{dx}{dt}\right)\frac{dx}{dt} + \omega^2 x = 0$$

is stable at the origin. $V = \omega^2 x^2 + (dx/dt)^2$ is a Liapunov function for this nonlinear equation.

EXAMPLE 15. LAGRANGE'S THEOREM ON THE STABILITY OF EQUILIBRIUM

Let $H = K + V$ be the Hamiltonian of a system with n degrees of freedom. Let $q_1, q_2, \ldots, q_n$ be n generalized coordinates, and $p_1, p_2, \ldots, p_n$ the corresponding n generalized momenta. The n equations of motion

$$\frac{d}{dt}\left(\frac{\partial K}{\partial p_i}\right) - \frac{\partial K}{\partial q_i} = -\frac{\partial V}{\partial q_i} \qquad i = 1, 2, \ldots, n \qquad (64)$$

together with n notational equations

$$\frac{dq_i}{dt} = p_i \qquad i = 1, 2, \ldots, n \tag{65}$$

form a $2n$th-order system as defined here. As is well known, (64) and (65) merely express the fact that the total t derivative of the Hamiltonian H along each trajectory vanishes, i.e.,

$$\frac{dH(\bar{x}(t), d\bar{x}/dt)}{dt} = \frac{d}{dt} H(q_1, q_2, \ldots, q_n, p_1, p_2, \ldots, p_n) = 0$$

Now the potential energy V depends solely upon the q_i and, typically, is analytic in the q_i, vanishes, say, at the origin and possesses a local minimum there. The kinetic energy K, typically, is a positive definite quadratic form in the p_i (i.e., vanishing only for $p_i = 0$, $i = 1, 2, \ldots, n$, and positive otherwise) with coefficients analytic in the q_i. Clearly, then, the Hamiltonian H is a Liapunov function for the system (64), (65) and the origin (static equilibrium position $q_1 = q_2 = \cdots = q_n = 0$) is stable.

♦ ♦ ♦ ♦ ♦ ♦ ♦

The Liapunov function may be useful for other than mere stability considerations. For example, it is clear that

$$\lim_{t \to \infty} V(\bar{x}(t)) = V_0 \geq 0$$

exists for each trajectory, passing sufficiently close to the origin at some time. If, in addition, $\bar{x}(t)$ is a nontrivial periodic solution, then $V(\bar{x}(t)) = V_0$, a constant greater than zero. The latter property is sometimes useful in locating periodic solutions or, as the case may be, in demonstrating the impossibility of periodic solutions. In Example 14, the only possible loci for periodic solutions are subsets of the ellipses

$$V(\bar{x}) = \omega^2 x^2 + \left(\frac{dx}{dt}\right)^2 = r^2$$

But these, obviously, do not contain solutions unless $k = 0$.

Geometrically we interpret the locus $V(\bar{x}) = $ constant as a surface enclosing the origin. It possesses the property that if any trajectory penetrates the surface, it does so from the outside to the inside. Of course, trajectories may also lie on the surface. We have seen that periodic solutions necessarily lie on such surfaces. More generally, the positive limit cycle of any solution $\bar{x}(t)$ (which passes sufficiently close to the origin at some time) must lie on a surface $V(\bar{x}) = $ constant. On the other hand, if the total t derivative $dV(\bar{x}(t))/dt$ is (and remains) actually negative, then the trajectory $\bar{x}(t)$ must actually penetrate each of the level surfaces ($V = $ constant) of V which it encounters. Thus we have Theorem 7.

THEOREM 7

If there exists a Liapunov function $V(\bar{x})$ for (62) so that in some neighborhood of the origin

$$\nabla V(\bar{x})\, \bar{f}(\bar{x}) < 0 \qquad (66)$$

for $\bar{x} \neq 0$, then (62) is asymptotically stable at the origin.

EXAMPLE 16

Consider Liénard's equation

$$\frac{d^2x}{dt^2} + f(x)\frac{dx}{dt} + g(x) = 0 \qquad (67)$$

and the equivalent system

$$\frac{dx}{dt} = y - F(x)$$
$$\frac{dy}{dt} = -g(x) \qquad (68)$$

where $F(x) = \int_0^x f(u)\, du$. We assume that $f(x)$ is a continuous function and $g(x)$ is a Lipschitz function, vanishing for $x = 0$. The Cauchy-Lipschitz existence-uniqueness theorems then apply to (67), and hence to (68). We note that the xy plane of (68) is not the usual phase plane associated with (67). The xy plane of (68) is referred to as the *Liénard plane*. If the inequalities

$$g(x) F(x) \geq 0$$
$$G(x) \geq 0 \tag{69}$$

where $G(x) = \int_0^x g(u) \, du$, hold in some neighborhood of $x = 0$, then

$$V(x,y) = \tfrac{1}{2}y^2 + G(x) \tag{70}$$

is a Liapunov function for (68). In fact

$$\nabla V(x,y) \, \tilde{f}(\tilde{x}) = (g(x), \, y) \begin{pmatrix} y - F(x) \\ -g(x) \end{pmatrix} = -g(x) F(x)$$

Clearly, if in some neighborhood of $x = 0$

$$g(x) F(x) > 0 \tag{71}$$

for $x \neq 0$, then (68) is asymptotically stable at the origin. Notice that the second inequality in (69) is guaranteed by the inequality $xg(x) \geq 0$ in some neighborhood of $x = 0$, i.e., if $g(x)$ assumes the sign of x for $|x|$ sufficiently small. In such a case, the first inequality in (69) is guaranteed by a similar condition on the sign of $f(x)$ for $|x|$ sufficiently small. Finally then, the inequality (71) is guaranteed by the two inequalities (largely fulfilled in applications)

$$xf(x) > 0$$
$$xg(x) > 0 \tag{72}$$

for $x \neq 0$ in a neighborhood of $x = 0$.

♦ ♦ ♦ ♦ ♦ ♦ ♦

For instability we have Theorem 8.

THEOREM 8

If in a neighborhood of the origin a scalar function $W(\tilde{x})$ is continuously differentiable, satisfies

$$\nabla W(\tilde{x}) \, \tilde{f}(\tilde{x}) > 0 \tag{73}$$

for $\bar{x} \neq 0$, and vanishes at the origin but is positive at some point of each neighborhood of the origin, then (62) is unstable at the origin.

PROOF: Let δ be a positive number such that the hypotheses apply for $|\bar{x}| \leq \delta$. Let M be the maximum of $W(\bar{x})$ for $|\bar{x}| \leq \delta$. Consider any positive number $\epsilon < \delta$. There exists a point $\bar{c}$, satisfying $|\bar{c}| \leq \epsilon$, such that $W(\bar{c}) > 0$. Since $W(\overline{0}) = 0$, there exists a positive number η such that $W(\bar{x}) < W(\bar{c})$ for $|\bar{x}| \leq \eta$. Clearly, $\eta < |\bar{c}| \leq \epsilon < \delta$. Let m be the minimum of $\nabla W(\bar{x}) \bar{f}(\bar{x})$ for $\eta \leq |\bar{x}| \leq \delta$. By (73), it is clear that $m > 0$. If $\bar{x}(t)$ is the trajectory with $\bar{c}$ as initial value for $t = 0$, then $dW(\bar{x}(t))/dt \geq 0$, so long as $|\bar{x}(t)| \leq \delta$. Hence $W(\bar{x}(t)) \geq W(\bar{c})$, and necessarily, $|\bar{x}(t)| \geq \eta$. Thus

$$\frac{dW(\bar{x}(t))}{dt} = \nabla W(\bar{x}(t)) \, \bar{f}(\bar{x}(t)) \geq m$$

so long as $|\bar{x}(t)| \leq \delta$. But this in turn implies that

$$W(\bar{x}(t)) \geq W(\bar{c}) + tm$$

and so either $W(\bar{x}(t))$ must eventually exceed M or $|\bar{x}(t)|$ eventually exceed δ. The latter is assured in either case since M is the maximum of $W(\bar{x})$ for $|\bar{x}| \leq \delta$. Thus from within each neighborhood of the origin stems a trajectory which penetrates the surface $|\bar{x}| = \delta$. The origin is manifestly unstable.

EXERCISES

1. Discuss the geometric meaning of (63). Consider the gradient ∇V as a vector normal to a level surface of V and $\bar{f}(\bar{x})$ as a direction field. What is a familiar name for the product in (63)?

2. Prove Theorem 7.

3. Explain why the second-order equation (67) is equivalent to the system (68), and construct a number of examples of Liénard's equation (67) which satisfy the inequalities (72).

4. Construct a number of examples illustrating Theorem 8.

5. State and prove theorems analogous to Theorems 6, 7, and 8 for nonautonomous systems.

6. Show that if $V(\bar{x})$ is a Liapunov function for (62), then the positive limit cycle of a solution passing sufficiently close to the origin lies on a surface $V(\bar{x}) = $ constant.

7. Consider the two-dimensional system

$$\frac{dx}{dt} = p(x,y)$$
$$\frac{dy}{dt} = q(x,y) \tag{i}$$

Assume that each of $p(x,y)$ and $q(x,y)$ is continuously differentiable and vanishes at the origin. Show that if the linear system

$$\frac{dx}{dt} = p_1 x + p_2 y$$
$$\frac{dy}{dt} = q_1 x + q_2 y \tag{ii}$$

where $p_1 = \partial p/\partial x$, $p_2 = \partial p/\partial y$, $q_1 = \partial q/\partial x$, and $q_2 = \partial q/\partial y$, each evaluated at the origin, is asymptotically stable at the origin, then

$$V(x,y) = (p_1 x + p_2 y)^2 + (q_1 x + q_2 y)^2 + \Delta(x^2 + y^2) \tag{iii}$$

where $\Delta = p_1 q_2 - q_1 p_2$, is a Liapunov function for (i). In fact, show that (ii) is asymptotically stable at the origin. Note first that necessarily $\Delta > 0$ and $p_1 + q_2 < 0$. (Why?) Then show that because $\Delta \neq 0$,

$$K(x,y) = (p_1 x + p_2 y)^2 + (q_1 x + q_2 y)^2 \geq m(x^2 + y^2) \tag{iv}$$

for a suitable positive constant m and all x and y. In fact, show that $K(x,y) = 0$ if and only if $x = 0$, $y = 0$ and hence $K(x,y) \geq m > 0$ on the unit circle $x^2 + y^2 = 1$. Then note that $K(\alpha x, \alpha y) = \alpha^2 K(x,y)$, and explain why (iv) applies throughout. Clearly, one has $V(x,y) \geq (m + \Delta)(x^2 + y^2)$ for all x and y. Finally, using the mean-value theorem, show that in a suitable neighborhood of the origin

$$\nabla V(x,y) \begin{pmatrix} p(x,y) \\ q(x,y) \end{pmatrix} < \frac{m}{2} (p_1 + q_2)(x^2 + y^2)$$

and explain why the last two inequalities justify the stated conclusions.

8. Discuss the relationship between Exercise 7 and Theorem 1.

9. With the notation of Exercise 7, show that along each non-trivial trajectory $x(t)$, $y(t)$ of (i), (sufficiently near the origin) the total t derivative of $V(x(t), y(t))$ satisfies the differential inequality

$$\frac{dV}{dt} < \frac{m}{2} \frac{p_1 + q_2}{\Delta + M} V$$

where M is a positive constant such that $K(x,y) \leq M(x^2 + y^2)$ holds throughout.   Using Gronwall's lemma, show that

$$\lim_{t \to \infty} V(x(t), y(t)) = 0$$

and explain the significance of this fact.

10. Describe the level curves of the scalar function $V(x,y)$ of (iii).   Note that for the linear system (ii) the scalar quantity $K(x,y)$ of (iv) is proportional to the kinetic energy while the scalar quantity $V(x,y)$ is proportional to the total energy.   Illustrate for the linear equation $d^2x/dt^2 + k\, dx/dt + \omega^2 x = 0$.

11. With the notation of Exercise 7, show that (i) is unstable at the origin if the real part of each of the characteristic numbers of (ii) is positive.   Note the signs of Δ and $p_1 + q_2$, and proceed as in Exercises 7 and 9.

12. With the notation of Exercise 7, show that (i) is unstable at the origin if exactly one of the characteristic numbers of (ii) is positive.   Introduce a suitable linear transformation

$$\begin{aligned} u_1 &= b_{11}x + b_{12}y \\ u_2 &= b_{21}x + b_{22}y \end{aligned} \tag{v}$$

so that the system (ii) is equivalent to a system of the form

$$\frac{du_1}{dt} = \lambda_1 u_1$$

$$\frac{du_2}{dt} = \lambda_2 u_2$$

(vi)

with, say, $\lambda_1 > 0$ and $\lambda_2 < 0$. Then show that certain solutions of the system obtained by applying the transformation (v) to (i) reflect the behavior of corresponding solutions of (vi) near the origin. Explain why this proves that (i) is unstable. Note that one has shown here that the origin is a saddle point for (i) if it is a saddle point for the linear approximation (ii).

13. Using the technique introduced in Exercise 12, show that the origin is a node or a focus for (i) if it is a node for the linear approximation (ii). Show that the origin is a center or a focus for (i) if it is a center for (ii). Show that the origin is a focus for (i) if it is a focus for (ii).

Chapter 6

TWO-DIMENSIONAL SYSTEMS

1. *Critical Points of Autonomous Systems*

The concept of a critical point is of particular importance in the study of the trajectories of two-dimensional autonomous systems. The very special (and somewhat restrictive) geometrical and topological properties of the Euclidean two-space are primarily responsible for this. Indeed, many of the results and concepts to be considered in this chapter are peculiarly two-dimensional and reflect geometrical or topological properties of two-space not shared by higher dimensional spaces. We shall consider, throughout, a system

$$\frac{dx}{dt} = p(x,y)$$
$$\frac{dy}{dt} = q(x,y)$$

$$(1)$$

in which the right-hand members are continuously differentiable in each variable. Thus, since (1) is autonomous, through each point of xy-space there passes one and only one trajectory. In particular, trajectories of (1) never cross or exhibit points of contact. This imposes severe restrictions on the geometric patterns. We note that a nontrivial trajectory corresponds to a periodic solution of (1) if and only if it passes through a single point at two different times.   Such a trajectory defines a *Jordan curve*, i.e., a simple closed curve which separates the xy plane into two disjoint, connected point sets.   Further, the independent variable t is a "sliding" parameter along each trajectory and generally needs to be specified only to the extent of indicating the direction of motion along a trajectory.   The direction of motion is reversed upon replacing t by $-t$, while the trajectory curves themselves remain unchanged.

The system (1) defines a direction field

$$\frac{dy}{dx} = \frac{q(x,y)}{p(x,y)} \tag{2}$$

which is indeterminate only at points where $q(x,y)$ and $p(x,y)$ vanish simultaneously.   Such points, of course, represent singular solutions (point solutions) of (1), but in the present context are referred to as *critical points*.   The latter name is used to emphasize the typically "critical" geometric properties of the direction field (2) near such points.   This is in contrast to the singular dynamical nature of the corresponding point solutions of (1).   The concepts are related but not equivalent.   For example, if $p(x,y)$ and $q(x,y)$ have a common factor which vanishes at some particular point, this point always represents a singular solution of the dynamical system (1).   On the other hand, the direction field defined by (2) may not be "critical" at all near such a point since a factor common to $p(x,y)$ and $q(x,y)$ cancels in (2).   Because it is the dynamical problem we are concerned with, the expression "critical point" will always refer to a point where $p(x,y)$ and $q(x,y)$ vanish simultaneously whether the direction field is actually critical or not.   We shall further assume that (1) possesses only isolated critical points.   Then the set of critical points will possess no finite limit points and any bounded region will contain at most a finite number of critical points.   A noncritical point will be called a regular point, as usual.   Generally, but not always, the topological (in fact, geo-

mctrical) characterization of the critical points of (1) are reflected
in the linear approximations of (1) near these critical points.   We
return to this question presently, but first introduce an interesting, often useful, concept.

Figure 1

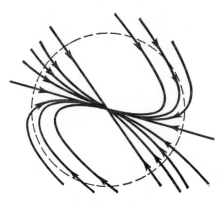

Figure 2

A significant property of a critical point P of (1) is its *index*.
This is an integer j which represents the net rotation (positive for
counterclockwise rotation) of the direction field along a simple
closed curve (a Jordan curve) J which is sufficiently smooth and
encloses the critical point P, but no other critical points.   The
index is a property of P and, therefore, the same for every such J.
In fact, it is expressed as the line integral

$$j = \frac{1}{2\pi} \oint_J d\left(\arctan \frac{q}{p}\right)$$

$$= \frac{1}{2\pi} \oint_J \frac{\left(p \dfrac{\partial q}{\partial x} - q \dfrac{\partial p}{\partial x}\right) dx + \left(p \dfrac{\partial q}{\partial y} - q \dfrac{\partial p}{\partial y}\right) dy}{p^2 + q^2} \qquad (3)$$

taken counterclockwise along J. This expression makes it evident that j varies continuously with any continuous deforma-

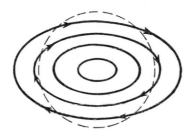

Figure 3

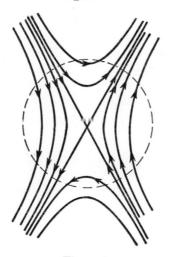

Figure 4

tions of J which do not lead to encounters with critical points. Since it is an integer, it is necessarily constant in the above circumstance.

Evidently, the index of a focus, node, or center is $+1$, while the index of a saddle point is -1 (see Figs. 1 to 4). We define the index of a regular point in an analogous manner and conclude immediately that it is zero (see Fig. 5).   More generally, one may define by (3), the index j for any simple closed curve J which is sufficiently smooth and does not pass through critical points.   Thus j is an integer valued function which is constant under any continuous deformations of J which do not lead to encounters with critical points.   The index of a simple closed curve is the sum of the indices of the enclosed critical points.   In particular, the index of a simple closed curve enclosing no critical points is zero.

Figure 5

If a simple closed curve J does not pass through any critical points and is, at the same time, a trajectory of (1), then clearly the index of J is $+1$ (see Fig. 3).   Such a trajectory corresponds to a nontrivial periodic solution of (1) and evidently must enclose at least one critical point.   We refer to such a trajectory as a *periodic orbit*.

We shall conclude this section with a result which implies, among other things, that the index of a critical point of (1) is generally the same as for its linear approximation.   Without loss of generality we may assume that the critical point is the origin.

Suppose $p(x,y)$ and $q(x,y)$ vanish simultaneously at the origin. Then we may write

$$\frac{dx}{dt} = p(x,y) = p_1 x + p_2 y + \rho(x,y)$$

$$\frac{dy}{dt} = q(x,y) = q_1 x + q_2 y + \eta(x,y)$$

(4)

where $p_1 = \partial p / \partial x$, $p_2 = \partial p / \partial y$, $q_1 = \partial q / \partial x$, and $q_2 = \partial q / \partial y$, each evaluated at the origin, and according to the mean-value theorem, conclude that

$$\lim_{[|x|+|y|]\to 0} \frac{|\rho(x,y)| + |\eta(x,y)|}{|x| + |y|} = 0$$

(5)

The latter is the appropriate nonlinearity condition for (4) near the origin, and it might be expected that near the origin the direction field defined by (4) is essentially given by the linear approximation

$$\frac{dx}{dt} = p^*(x,y) = p_1 x + p_2 y$$

$$\frac{dy}{dt} = q^*(x,y) = q_1 x + q_2 y$$

(6)

The expected does in fact occur if (6) is not degenerate, that is, if the origin is an isolated critical point of (6). This is equivalent to the nonvanishing of the determinant of the coefficients. We obtain the desired result by considering the dot product of the vector (p,q) defined by (4) with the vector (p^*,q^*) defined by (6). This is the scalar quantity

$$D = (p_1 x + p_2 y)^2 + (q_1 x + q_2 y)^2 + (p_1 x + p_2 y)\rho(x,y)$$
$$+ (q_1 x + q_2 y)\eta(x,y) \quad (7)$$

We shall divide (7) by the product of the lengths of the two vectors; the quotient becomes the cosine of the angle θ between the two vectors (see Fig. 6). On the other hand, if the determinant

$$\begin{vmatrix} p_1 & p_2 \\ q_1 & q_2 \end{vmatrix} = p_1q_2 - q_1p_2$$

is not zero, the two quantities $p_1x + p_2y$ and $q_1x + q_2y$ vanish simultaneously only at the origin. Thus the quantity

$$K(x,y) = (p_1x + p_2y)^2 + (q_1x + q_2y)^2$$

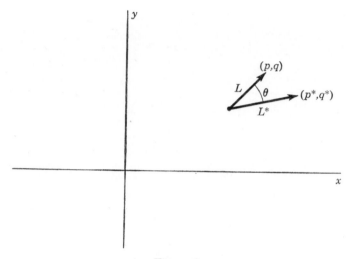

Figure 6

does not vanish on the square $|x| + |y| = 1$. In particular, there exists a positive constant m such that

$$K(x,y) \geq m \qquad \text{for} \qquad |x| + |y| = 1$$

Clearly, then, $K(x,y) \geq m(|x| + |y|)^2$ for *all* (x,y). The lengths of the two vectors are given by the expressions

$$L = [K(x,y)]^{\frac{1}{2}} \left[1 + \frac{2(p_1x + p_2y)\rho(x,y) + 2(q_1x + q_2y)\eta(x,y) + \rho^2(x,y) + \eta^2(x,y)}{K(x,y)} \right]^{\frac{1}{2}}$$

and

$$L^* = [K(x,y)]^{\frac{1}{2}}$$

Hence from (7), we obtain

$$\cos \theta = \frac{D}{LL^*} = \frac{1 + \dfrac{(p_1 x + p_2 y)\rho(x,y) + (q_1 x + q_2 y)\eta(x,y)}{K(x,y)}}{1 + \left[\dfrac{2(p_1 x + p_2 y)\rho(x,y) + 2(q_1 x + q_2 y)\eta(x,y) + \rho^2(x,y) + \eta^2(x,y)}{K(x,y)}\right]^{1/2}}$$

(8)

where

$$\left| \frac{2(p_1 x + p_2 y)\rho(x,y) + 2(q_1 x + q_2 y)\eta(x,y) + \rho^2(x,y) + \eta^2(x,y)}{K(x,y)} \right|$$

$$\leq \frac{2}{m^{1/2}} \frac{|\rho(x,y)| + |\eta(x,y)|}{|x| + |y|} + \frac{1}{m}\left[\frac{|\rho(x,y)| + |\eta(x,y)|}{|x| + |y|}\right]^2 \quad (9)$$

and

$$\left| \frac{(p_1 x + p_2 y)\rho(x,y) + (q_1 x + q_2 y)\eta(x,y)}{K(x,y)} \right| \leq \frac{1}{m^{1/2}} \frac{|\rho(x,y)| + |\eta(x,y)|}{|x| + |y|}$$

(10)

The nonlinearity condition (5) thus guarantees that the direction fields of (4) and (6) are nearly coincident near the origin; i.e., the angle θ between corresponding vectors of the two vector fields is *uniformly* small in a complete neighborhood of the origin. Clearly, this implies that the index of the origin for (4) is the same as for the linear approximation (6). But even more, we have essentially proved Theorem 1.

THEOREM 1

If in the system (4), the coefficients of the explicit linear terms satisfy $p_1 q_2 - q_1 p_2 \neq 0$ and the perturbational terms $\rho(x,y)$ and $\eta(x,y)$ are continuous and satisfy the nonlinearity condition (5), then the character of the critical point $x = 0$, $y = 0$ is the same as for the linear approximation (6) with one possible exception. If the critical point is a center for (6) it may be either a center or a focus for (4).

PROOF: In view of the above discussion, the only questionable cases are those involving either foci or centers. But the stability theory of Chap. 5 guarantees that a focus for (6) is a focus for (4) and that the stability of the focus is the same in each case. Thus, the stated exception is indeed the only possible exception. If the origin is a center for (6), it may also be one for (4). However, the inequalities (9) and (10) do not rule out the possibility that it might be a focus for (4).

We refer to centers, foci, nodes, and saddle points as *elementary* critical points. Theorem 1 shows that these are the types of critical points that are most likely to occur.

EXERCISES

1. Using (3), verify directly that the index of a focus, node, or center is $+1$, the index of a saddle point is -1, the index of a regular point is zero, and the index of a nonsingular closed trajectory is $+1$.

2. Show that the index of a simple closed curve in the continuous vector field defined by (1) is the sum of the indices of the enclosed critical points.

3. Determine the index of the critical point of the system

$$\frac{dx}{dt} = 2xy, \qquad \frac{dy}{dt} = y^2 - x^2$$

Note that the critical point is not a focus, a node, a center, nor a saddle point. What is the linear approximation to this system near the critical point?

4. Supposing that (1) possesses only elementary critical points, explain why a periodic orbit of (1) necessarily encloses an odd number of critical points. Explain why it is that if such a trajectory encloses more than one critical point, not all the enclosed critical points can be stable and not all the enclosed critical points can be unstable.

5. Give some general conditions under which (1) possesses only elementary critical points.

6. Give a detailed proof of Theorem 1.

7. Show that a trajectory is a periodic orbit of (1) if and only if it passes through a single point at two different times.   Explain why it is necessarily a Jordan curve.

2. *Properties of Limit Cycles*

We recall that a point P belongs to the positive limit cycle of a trajectory Γ if the trajectory intersects each neighborhood of P for arbitrarily large values of t, the independent variable.   The positive limit cycle of Γ is always a subset (possibly empty) of the set of limit points of each positive half path associated with Γ.   For autonomous systems, as considered here, each half path of a trajectory Γ is characterized as that portion of Γ extending to one side of a point of Γ.   Since the set of limit points of a half path is a closed point set, so also is the intersection of any number of such sets.   In particular, a limit cycle is always a closed point set.   Further, if a positive half path is bounded, then the positive limit cycle is a nonempty closed and bounded point set, i.e., a compact set.   It may also be shown that the positive limit cycle is always a connected point set.

The following theorems depict other important properties of limit cycles.

THEOREM 2

If a positive limit cycle L of a trajectory Γ of (1) contains a regular point P, then the complete trajectory Γ^* through P is contained in L.   If in addition L is bounded, then Γ^* exists for all t, and the positive and negative limit cycles of Γ^* are each contained in L.

PROOF: Let $P_1 \neq P$ be a point of Γ, N_1 a neighborhood of P_1, and T_1 a real number.   According to the basic continuity theorem of Chap. 3, there exists a neighborhood N of P such that each trajectory intersecting N also intersects N_1.   If the trajectory Γ moves from P to P_1 with increasing t, then this is a direct application of the continuity theorem. If Γ moves from P_1 to P with increasing t, then the continuity theorem is applied to (1) with t replaced by $-t$. In the latter circumstance, let τ be an upper bound to the elapse time for travel from N_1 to N along any trajectory intersecting N_1. The existence of the latter is also guaranteed by the continuity theorem. Since P is contained in the limit cycle L, there exists a time t_1 exceeding T_1 (or $T_1 + \tau$, as the case may be) such that Γ intersects N for $t = t_1$. Then Γ necessarily intersects N_1 for some $t_2 \geq T_1$. This shows that

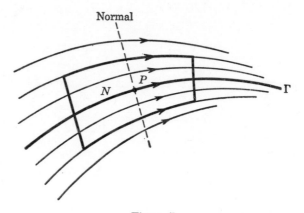

Figure 7

P_1 is also contained in the positive limit cycle of Γ. If the limit cycle L is contained in a bounded region, then a standard continuation argument shows that the trajectory Γ^* through P exists for all t and is contained in L. Since L is closed, the positive and negative limit cycles of Γ^* are necessarily contained in L.

By arguments similar to those used in the proof of Theorem 2, one may show that if P is a regular point of the system (1) and Γ is the trajectory through P, then there exists a neighborhood N of P such that each trajectory intersecting N also intersects the normal to Γ at P. This is essentially a manifestation of the uniform continuity of the direction field near P. We may further require that each trajectory in N cross the normal in the same direction as does Γ.   One thinks of a locally laminar flow field near P (see Fig. 7).   The segment of the normal at P lying within N is called a *transversal*, and the neighborhood N is called a *path-rectangle*.

THEOREM 3

If a bounded positive half path of a trajectory Γ of (1) and its positive limit cycle L have a regular point P in common, then $\Gamma = L$ and L is a periodic orbit.

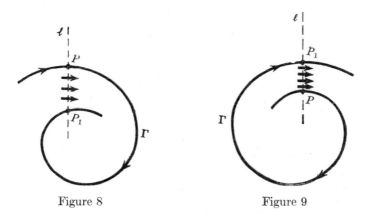

Figure 8   Figure 9

PROOF: According to Theorem 2, the complete trajectory Γ through P is contained in L.   Let us show that Γ is a periodic orbit.   In fact, let l be a transversal at P, and N the corresponding path-rectangle.   Let the trajectory Γ arrive at P for $t = t_1$. Then, since P belongs to the limit cycle L, the trajectory Γ must intersect the neighborhood N, and thence also the transversal l, for $t > t_1$.   If a second intersection with l is again P, then clearly Γ is periodic.   On the other hand, if t_2 is the least time exceeding t_1 for which Γ intersects l at a point P_1 different from P, then that portion of the trajectory Γ mapped out for $t_1 \leq t \leq t_2$, together with the segment of the transversal l between P and P_1, defines a Jordan curve J (see Figs. 8 and 9).   The trajectory Γ for

$t > t_0$ lies either completely on the outside of J or completely on the inside of J since it cannot cross itself nor the transversal between P and P_1. In either event, the trajectory cannot return to any small path-rectangle at P which excludes P_1. In such a case, P could not be a point of the limit cycle L. Since P *is* a point of the limit cycle L, we conclude that the second intersection of Γ with the transversal γ is necessarily P itself, and hence that Γ is a periodic orbit. Clearly, the trajectory Γ and its limit cycle L correspond to one and the same *closed* point set.

A very similar argument may be employed to prove Theorem 4.

THEOREM 4

If a trajectory of (1) intersects a transversal at two points, then the trajectory is not contained in any limit cycle of (1).

PROOF: Consider the Jordan curve defined by that portion of a trajectory Γ mapped out between two successive intersections with a transversal and the corresponding segment of the transversal. We observe that a trajectory can pass either from the inside of this Jordan curve to the outside or from the outside to the inside at most once, since it must do so by crossing the transversal (see Figs. 8 and 9). Clearly not both points of intersection of Γ with the transversal can belong to one and the same limit cycle.

We now consider the celebrated Poincaré-Bendixson theorem.

THEOREM 5

Let a trajectory Γ of (1) possess a bounded positive half path. If the positive limit cycle L of Γ consists of regular points only, then L is a periodic orbit and either $\Gamma = L$ or Γ "spirals" to L on one side of L.

PROOF: The hypotheses imply that L is not an empty set. Let P be a point of L. Since it is necessarily a regular point, Theorem 2 guarantees that the complete trajectory Γ^* through P is contained in L. The positive limit cycle L^* of Γ^* is not empty and is also contained in L.   Let P^* be a point in L^*.   Now P^* is a regular point and according to Theorem 3, Γ^* can intersect the transversal at P^* in no more than one point, since Γ^* is a subset of the limit cycle L.   Clearly, Γ^* must intersect the transversal in every neighborhood of P^*, since P^* is, at the same time, in the limit cycle L^* of Γ^*.   Thus Γ^* must intersect the transversal at P^* and only at P^*.   But then according to Theorem 3, $\Gamma^* = L^*$ is a periodic orbit.   Since the limit cycle L is connected, it cannot contain points other than those of the closed

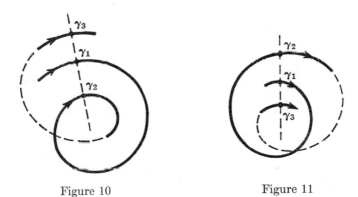

Figure 10   Figure 11

periodic orbit $\Gamma^* = L^*$.   Thus L is a periodic orbit as stated. Of the two possibilities for Γ, if Γ has a point in common with L, then $\Gamma = L$.   If Γ has no points in common with L, let P be any point of L.   If l is a transversal at P, then Γ intersects l consecutively in a succession of points $\gamma_1, \gamma_2, \ldots$ on one side of L with $\lim_{k \to \infty} \gamma_k = P$.   In fact, the consecutive intersections of any trajectory with a transversal necessarily form a monotone sequence along that transversal (see Figs. 10 and 11).   Clearly, these must lie on a single side of L in the present circumstance.

Finally, we characterize limit cycles in general according to Theorem 6.

THEOREM 6

Let a trajectory Γ of (1) possess a bounded positive half path. Then the positive limit cycle L of Γ is one of the following:

a. A single critical point of (1)

b. A periodic orbit

c. A finite number of critical points of (1) with connecting trajectories (nonperiodic orbits) called separatrices, each of which has for its limit cycle (positive or negative), one of the critical points

PROOF: Clearly each of a and b is a possibility. The hypotheses guarantee that the limit cycle L is not an empty set. According to Theorem 3, a limit cycle cannot contain both critical points and periodic orbits. If it contains no singular points, Theorem 5 implies b. If it contains critical points, they are but finite in number. Since a limit cycle is necessarily connected, if it contains more than one critical point these must be connected by regular points and hence, according to Theorem 2, by complete trajectories. Clearly, the positive and negative limit cycles of the connecting trajectories are the critical points contained in L. Thus, c is the only alternative to either one of a or b.

EXAMPLE 1

Theorem 5 implies that if a bounded region R contains a half path but no critical points, then it also contains a periodic orbit. One can be assured that a bounded region R contains a half path if its boundary consists of two concentric Jordan curves and the vector field along the boundary is everywhere directed into (or everywhere directed outward from) the annular region between the two Jordan curves (see Fig. 12).

EXAMPLE 2

If a Jordan curve encloses exactly one unstable critical point and the vector field along this curve is everywhere directed into the interior, then there is a periodic orbit lying within the Jordan curve.   Of course any periodic orbit lying within the Jordan curve must encircle the critical point.   Thus if there is more than one, the orbits must be concentric.   If there is just one, it is necessarily asymptotically orbitally stable (see Fig. 13).   In fact, each

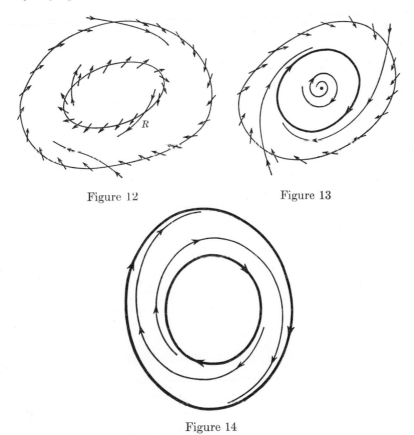

Figure 12   Figure 13

Figure 14

trajectory crossing the Jordan curve must spiral in to the periodic orbit, and each nontrivial trajectory within the periodic orbit must spiral out to the periodic orbit.   There are no other possibilities for the positive limit cycles of such trajectories.   For exactly the same reason, if there are no critical points or periodic orbits between two concentric periodic orbits, such as in Fig. 14, then each trajectory between the periodic orbits has one of the periodic orbits for its positive limit cycle and the other for its negative limit cycle, i.e., the interior trajectories depict motion from one of the periodic orbits to the other for $-\infty < t < +\infty$.

EXERCISES

1. Explain why the limit cycle of a bounded half path is nonempty.

2. Show that a limit cycle is a connected point set. In fact, show that if a limit cycle is the union of two disjoint, nonempty, closed sets, then a relevant half path could not be connected. Note that two disjoint, nonempty, closed sets are at a nonzero distance from each other.

3. Prove the existence of a transversal and a corresponding path-rectangle at a regular point.

4. Give a detailed proof of Theorem 4.

5. State and prove results analogous to Theorems 2, 3, 4, 5, and 6 for negative limit cycles.

6. Justify the statements appearing in Example 1.

7. Extend the results in Example 1 to cases wherein the vector field might be tangent to portions of the relevant Jordan curves. Thus, in some cases, portions of the Jordan curves might be trajectories. Use this and the result in Example 1 to prove the principal results in Example 2.

8. Consider the system

$$\frac{dx}{dt} = y + xp(x^2 + y^2)$$
$$\frac{dy}{dt} = -x + yp(x^2 + y^2)$$

(i)

where p is continuous for all positive numbers. Show that (i) leads to the single equation

$$\frac{dr^2}{dt} = 2r^2p(r^2)$$

(ii)

for $r^2 = x^2 + y^2$. Explain why the origin is the only critical point of (i). Suppose that the function p has a positive root r_0^2. Then $r^2 = r_0^2$ is a singular solution of (ii) and $x^2 + y^2 = r_0^2$ is a periodic orbit of (i). Suppose, further, that p possesses only a finite number of positive roots. Characterize the stability characteristics of the origin and each of the periodic orbits of (i), corresponding to the positive roots of p, in terms of the sign of p between the roots.

9. Discuss the trajectories of the system

$$\frac{dx}{dt} = y + x(R^2 - x^2 - y^2)$$

$$\frac{dy}{dt} = -x + y(R^2 - x^2 - y^2)$$

for various positive constants R^2.

10. Consider the divergence

$$\text{div } (p,q) - \frac{\partial p}{\partial x} + \frac{\partial q}{\partial y}$$

of the vector field defined by (1). As the name implies, the quantity div (p,q) at a point is a measure of the extent to which the vectors near that point are diverging (separating). A negative divergence means that the vectors are converging. Therefore, the sign of div (p,q) along a trajectory is an indicator of stability. If div (p,q) is negative everywhere along a trajectory, then the trajectory would appear to be orbitally stable. For a periodic orbit, one might expect a little more. The neighboring trajectories of a periodic trajectory must accompany (approximately, of course) the periodic one, and so it would appear that the stability should be indicated merely by the sign of the *average* of the div (p,q) along a periodic trajectory. Explain the relationship between this "intuitive" result and the Poincaré criterion for orbital stability discussed in Example 12 of Chap. 5.

11. Using Green's theorem, show that the total (and hence the space average of the) divergence of the vector field (p,q) within a periodic orbit of (1) must vanish. Explain why this is "intuitively" necessary. What would a positive or negative total or average divergence signify?

12. Consider the van der Pol equation

$$\frac{d^2x}{dt^2} + \mu(x^2 - 1)\frac{dx}{dt} + x = 0 \qquad \text{(i)}$$

where μ is a positive constant. One may use the Poincaré-Bendixson theorem, Theorem 5, and show that (i) possesses a periodic orbit. First show that (i) is equivalent to the system

$$\frac{dx}{dt} = y + \mu\left(x - \frac{x^3}{3}\right)$$
$$\frac{dy}{dt} = -x \qquad \text{(ii)}$$

Note that this is not the usual phase-plane equivalent of (i). Show that the origin is the only critical point of (ii) and that it is unstable. The direction field defined by (ii) may be expressed in the form

$$\frac{dy}{dx} = -\frac{x}{y + \mu(x - x^3/3)} \qquad \text{(iii)}$$

Thus, in particular, a trajectory becomes horizontal only as it crosses the y axis and becomes vertical only as it crosses the cubic

$$y = \mu\left(\frac{x^3}{3} - x\right) \qquad \text{(iv)}$$

The cubic curve (iv) is illustrated in Fig. 15. Explain why the trajectories of (ii) are symmetrical with respect to the origin. Then using (iii), show that each nontrivial trajectory Γ which intersects the cubic curve (iv) for $x > \sqrt{3}$ is qualitatively as shown in Fig. 15. Next observe that the quantity $E = x^2 + y^2$

(proportional to the energy) satisfies the differential equation

$$\frac{dE}{dt} = -2\mu x \left(\frac{x^3}{3} - x \right) \qquad \text{(v)}$$

along each trajectory of (ii).   Thus, E decreases along any portion of Γ lying to the right of $x = \sqrt{3}$.   In fact, from the second

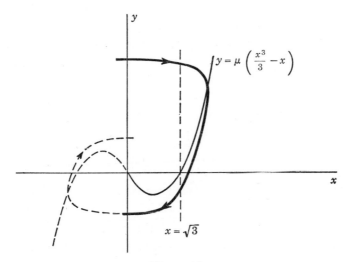

Figure 15

equation in (ii) and (v), one has

$$\frac{dE}{dy} = 2\mu \left(\frac{x^3}{3} - x \right)$$

and so the decrease in E along any portion of Γ to the right of
$x = \sqrt{3}$ will numerically exceed the change in y for that portion
multiplied by the minimum value of $2\mu(x^3/3 - x)$ over that
portion.   Explain why the decrease in E along the totality of Γ
lying to the right of $x = \sqrt{3}$ is numerically unbounded as the
intercept of Γ with the cubic (iv) moves to the right.   On the

other hand, from the first equation in (ii) and (v), one has

$$\frac{dE}{dx} = -\frac{2\mu x \left(\frac{x^3}{3} - x\right)}{y + \mu \left(x - \frac{x^3}{3}\right)} \tag{vi}$$

and so the increase in E along any portion of Γ lying between $x = 0$ and $x = \sqrt{3}$ does not numerically exceed the change in x multiplied by the maximum of the right-hand member in (vi). Explain why the increase in E along each of the two portions of Γ lying between $x = 0$ and $x = \sqrt{3}$ is bounded and, in fact, tends to zero as the intercept of Γ with the cubic (iv) moves to the right. Now explain why the y intercept of the trajectory Γ above the x axis is farther from the origin than is the y intercept of Γ below the x axis, provided the intercept of Γ with the cubic (iv) is sufficiently far to the right. Finally, use a loop of such a trajectory Γ and its reflection in the origin, together with suitable segments of the y axis, and define a Jordan curve J which encloses the origin and for which the direction field is everywhere either tangent to J or directed into the interior of J. Explain why this shows that (ii), and hence (i), possesses a periodic orbit.

13. By arguments similar to those used in Exercise 11, show that a trajectory Γ of the van der Pol system (ii) which intercepts the cubic (iv) between $x = 0$ and $x = \sqrt{3}$ possesses a y intercept above the x axis which is nearer the origin than is its succeeding y intercept below the origin. Show therefore that the periodic orbit of (i) is unique. Why is it necessarily asymptotically orbitally stable?

14. Discuss the trajectories of the van der Pol equation (i) for $\mu < 0$. Note that changing μ to $-\mu$ in (i) is equivalent to changing t to $-t$.

15. For $|\mu|$ small, the periodic orbit of the van der Pol equation (i) is very nearly a circle. This becomes clear if one examines the direction field of the equivalent system (ii). This is the

Liénard plane associated with (i).   As $|\mu|$ decreases, the illustration in Fig. 15 is modified so as to move the cubic curve (vi) closer to the x axis, as in Fig. 16.   For $\mu = 0$, the cubic passes to the x axis.

The direction field at a point (x,y) in Fig. 16 is given by (iii) and, as indicated, is perpendicular to a ray drawn from a point P on the y axis only slightly removed from the origin.   Explain why this is a valid geometric interpretation of (iii).   If the ray were always drawn from the origin, the trajectory would be a

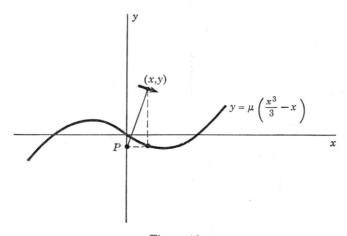

Figure 16

circle with the origin as center.   Because of the slight movement of P away from the origin as one progresses along a trajectory, the trajectory itself deviates slightly from a circle.   Using this construction (the *Liénard construction*), explain why it is that the trajectories near the origin necessarily spiral outward, while those at some distance from the origin necessarily spiral inward. The periodic orbit is that unique trajectory along which the two effects, during one rotation, exactly balance.

16. Using the Liénard construction, discuss the trajectories of the van der Pol system (ii) for large μ.   For this discussion, replace the independent variable t by $\tau = \mu t$ and the dependent

variable y by $z = y/\mu$. Note the limiting form for the periodic orbit in the xz plane as $\mu \to \infty$. Show that the corresponding motion for (i) is a very jerky type of motion. This is called *relaxation oscillation*. The ordinary phase-plane trajectories of (i) for μ large are illustrated in Fig. 17. By plotting isoclines

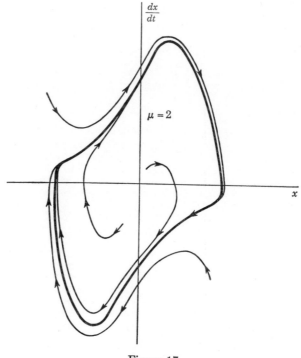

Figure 17

or by other geometric methods, show that this illustration is qualitatively correct.

17. Consider an equation of the form

$$\frac{d^2x}{dt^2} + f(x)\frac{dx}{dt} + g(x) = 0$$

Introduce suitable assumptions concerning the functions $f(x)$ and $g(x)$ so that the essential arguments used in the discussion of the van der Pol equation carry over to a general family of similar equations.

18. Rephrase the results of Exercise 17 for the system

$$\frac{d^2x}{dt^2} + F\left(\frac{dx}{dt}\right) + G(x) = 0 \qquad \text{(vii)}$$

noting that if F and G are each continuously differentiable, then a solution of (vii) also satisfies the equation

$$\frac{d^3x}{dt^3} + F'\left(\frac{dx}{dt}\right)\frac{d^2x}{dt^2} + G'(x)\frac{dx}{dt} = 0 \qquad \text{(viii)}$$

Consider, in particular, the case $G(x) = x$ and the ordinary phase-plane trajectories of (vii). Explain the relationship between these and the Liénard plane trajectories of (viii). Illustrate for the van der Pol case.

19. Introduce a Liénard-type construction for equations of the form $d^2x/dt^2 + F(dx/dt) + x = 0$, using a geometric interpretation of the direction field

$$\frac{dy}{dx} = -\frac{x + F(y)}{y}$$

in the usual phase plane. Consider the locus $x = F(y)$ and an appropriate ray originating along the x axis.

20. Apply the Liénard construction of Exercise 19 to the equation $d^2x/dt^2 \pm k + \omega^2 x = 0$, where the plus sign prevails whenever $dx/dt > 0$, and the minus sign whenever $dx/dt < 0$. This type of damping is called *Coulomb damping* (or sliding friction). Integrate the equation in closed form, and discuss the relationship between the closed form solutions and the geometric solutions in the phase plane.

Chapter 7

PERTURBATIONS OF PERIODIC SOLUTIONS

1. *Perturbations of Periodic Solutions in Nonautonomous Systems*

We continue here the quest for, and study of, periodic solutions begun in the previous chapter. A number of results will be presented for rather general n-dimensional systems and a number for special two-dimensional systems. However, the latter are *not* peculiarly two-dimensional necessarily, as were most of the results of the previous chapter. Generally, the two-dimensional version is used for ease of expression or for greater practical emphasis. Also, in contrast to the previous chapter, we shall here be concerned with nonautonomous as well as autonomous systems. As implied by the chapter title, all the results are of a perturbational character.

In a typical circumstance, one is concerned with the forced oscillations of a nonlinear system in which the forcing function is periodic. The study of the possible periodic responses of the same period may well be of first importance. The system may be one of a class of systems for which one member of the class is either very simple (it may be linear, for example) or for some reason or other has been previously investigated. If the one system possesses a periodic response, a number of rather natural questions arise: Do the neighboring systems also possess periodic responses?   Is the one periodic response a member of a continuous (or possibly analytic) family of periodic responses?   Are the stability properties of the one periodic response reflected in neighboring periodic responses?

Consider, for example, a system

$$\frac{d\bar{x}}{dt} = \bar{f}(\bar{x},t,\beta) \tag{1}$$

where β is a real (scalar) parameter, and where the right-hand member is continuously differentiable in each variable and is periodic in t of period $\tau > 0$.   Suppose that for $\beta = 0$, (1) possesses a periodic solution $\bar{p}(t)$ of period τ.   The stability of this periodic solution may be reflected in the equation of first variation

$$\frac{d\bar{y}}{dt} = \bar{f}_{\bar{x}}(\bar{p}(t), t, 0)\bar{y} \tag{2}$$

In fact, if the trivial solution of (2) is asymptotically stable, then the periodic solution $\bar{p}(t)$ is asymptotically stable.   It is intriguing to find that asymptotic stability of the origin in (2) also guarantees that for each sufficiently small value of $|\beta|$, Eq. (1) possesses a periodic solution of period τ.   More generally, we have Theorem 1.

THEOREM 1

If the equation of first variation (2) has no solution of period τ, (other than $\bar{y} = \bar{0}$) then for each $|\beta|$ sufficiently small, (1) possesses a (unique) solution $\bar{x}(t,\beta)$ which is periodic in t of period τ and continuous in the pair (t,β) with $\lim_{\beta \to 0} \bar{x}(t,\beta) = \bar{p}(t)$, uniformly for all t.

Proof: The idea of the proof is to imbed the periodic solutions $\bar{x}(t,\beta)$ in a much larger family of solutions.   Rather, we consider a family of (not necessarily periodic) solutions of sufficient generality that we are able to discover the desired periodic solutions imbedded within this family. To be specific, let $\bar{x}(t,\beta,\bar{c})$ denote the (unique) solution of (1) satisfying the initial condition

$$\bar{x}(0,\beta,\bar{c}) = \bar{p}(0) + \bar{c} \tag{3}$$

The hypotheses imply that for each β and $\bar{c}$ the solution $\bar{x}(t,\beta,\bar{c})$ exists, at least for a positive interval of time. For $|\beta|$ and $|\bar{c}|$ small, the solution $\bar{x}(t,\beta,\bar{c})$ approximates the periodic solution $\bar{p}(t) = \bar{x}(t,0,\bar{0})$. In particular, we shall always assume that $|\beta|$ and $|\bar{c}|$ are so small that the solution $\bar{x}(t,\beta,\bar{c})$ exists, say for $0 \le t \le 2\tau$. Since the system (1) is periodic in t of period τ, the solution $\bar{x}(t,\beta,\bar{c})$ will be periodic with period τ if

$$\bar{x}(\tau,\beta,\bar{c}) = \bar{x}(0,\beta,\bar{c})$$

or, what is the same, if

$$\bar{x}(\tau,\beta,\bar{c}) - \bar{p}(0) - \bar{c} = \bar{0} \tag{4}$$

We may regard (4) as an algebraic equation in $\bar{c}$ and β. Clearly, for $\beta = 0$, (4) possesses the solution $\bar{c} = \bar{0}$, corresponding to the periodic solution $\bar{p}(t) = \bar{x}(t,0,\bar{0})$. Thus, if the appropriate Jacobian [determinant of the derivative with respect to $\bar{c}$ of the left-hand member in (4)] does not vanish for $\beta = 0$, then the implicit function theorem guarantees that for each sufficiently small $|\beta|$, (4) possesses a (unique) solution $\bar{c} = \bar{c}(\beta)$ which is continuous in β (in fact, continuously differentiable in β) with $\lim_{\beta \to 0} \bar{c} = \bar{0}$. This means that for each sufficiently small $|\beta|$, there exists a (unique) choice of initial vector $\bar{c}(\beta)$ so that the particular member

$$\bar{x}(t,\beta,\ \bar{c}(\beta)) = \bar{x}(t,\beta)$$

of the general family is a periodic solution of (1) of period τ. Clearly $\lim_{\beta \to 0} \bar{x}(t,\beta,\ \bar{c}(\beta)) = \bar{p}(t)$, uniformly for *all* t, since the convergence is uniform over one period τ. On the other hand, the referenced Jacobian is the determinant

$$J = |\bar{x}_{\bar{c}}(\tau,0,\bar{0}) - I| \tag{5}$$

where I is the identity matrix and $\bar{x}_{\bar{c}}(\tau,0,\bar{0})$ is the Jacobian matrix

$$\bar{x}_{\bar{c}} = \begin{vmatrix} \dfrac{\partial x_1}{\partial c_1} & \dfrac{\partial x_1}{\partial c_2} & \cdots & \dfrac{\partial x_1}{\partial c_n} \\[2mm] \dfrac{\partial x_2}{\partial c_1} & \dfrac{\partial x_2}{\partial c_2} & \cdots & \dfrac{\partial x_2}{\partial c_n} \\[2mm] \cdots & \cdots & \cdots & \cdots \\[2mm] \dfrac{\partial x_n}{\partial c_1} & \dfrac{\partial x_n}{\partial c_2} & \cdots & \dfrac{\partial x_n}{\partial c_n} \end{vmatrix} \tag{6}$$

evaluated for $\beta = 0$, $\bar{c} = \bar{0}$, and $t = \tau$. If we consider for the moment the Jacobian matrix (6) and the general family of solutions $\bar{x}(t,\beta,\bar{c})$ of (1) for arbitrary (independent) β and $\bar{c}$, then we observe (by differentiation in (1)) that $X = \bar{x}_{\bar{c}}$ is a solution of the *matrix* equation of first variation

$$\frac{dX}{dt} = \bar{f}_{\bar{x}}(\bar{x}(t,\beta,\bar{c}), t, \beta)X \tag{7}$$

For $\beta = 0$ and $\bar{c} = \bar{0}$, (7) becomes

$$\frac{dX}{dt} = \bar{f}_{\bar{x}}(\bar{p}(t), t, 0)X \tag{8}$$

which is the matrix equation associated with the vector equation of first variation (2). Thus the Jacobian $\bar{x}_{\bar{c}}$ in (6), evaluated for $\beta = 0$ and $\bar{c} = \bar{0}$, is a matrix solution $X = \bar{x}_{\bar{c}}$ of (8). But, by differentiation in (3), it follows that $\bar{x}_{\bar{c}}(0,\beta,\bar{c}) = I$ (the identity matrix) and so, in particular, that $\bar{x}_{\bar{c}}(0,0,\bar{0}) = I$. Hence $\bar{x}_{\bar{c}}(t,0,\bar{0})$ is the *principal* matrix solution $Y(t)$ of (8). The determinant (5) may be reformulated as

$$J = |Y(\tau) - I| \tag{9}$$

Now (9) is zero if and only if the matrix $Y(\tau)$ has the number one for a characteristic number (eigenvalue). In this circumstance, the characteristic numbers are referred to as characteristic multipliers. What is significant here is the fact that the vector equation of first variation (2) possesses a nontrivial periodic solution of period τ if and only if one of the characteristic multipliers is unity. Thus, in particular, if the vector equation of first variation (2) has no periodic solutions of period τ (other than $\bar{y} = \bar{0}$), then the determinant J in (9) cannot be zero and hence the (unique) family of periodic solutions $\bar{x}(t,\beta)$ exists, as shown above.

EXAMPLE 1

Consider the linear scalar equation $dx/dt + \beta x = \cos t$. For $\beta = 0$, $x = \sin t$ is a periodic solution of period 2π. The equation of first variation is $dy/dt + \beta y = 0$ and, for $\beta = 0$, certainly does not admit a periodic solution of period 2π other than $y = 0$. Thus there exists a (unique) family of periodic solutions of period 2π for small $|\beta|$ converging to $x = \sin t$ as $\beta \to 0$. This is the family $x = [\beta/(1 + \beta^2)] \cos t + [1/(1 + \beta^2)] \sin t$.

EXAMPLE 2

The harmonically forced Duffing equation

$$\frac{d^2x}{dt^2} + \omega^2 x + \beta x^3 = F \cos \lambda t$$

possesses, for $\beta = 0$, the periodic solution $x = [F/(\omega^2 - \lambda^2)] \cos \lambda t$ of period $2\pi/\lambda$, assuming $|\lambda| \neq |\omega|$. The equation of first variation is $d^2y/dt^2 + (\omega^2 + 3\beta x^2)y = 0$ which, for $\beta = 0$, becomes $d^2y/dt^2 + \omega^2 y = 0$. This equation admits only nontrivial periodic solutions of the periods $\pm 2\pi/\omega$, $\pm 4\pi/\omega$, $\pm 6\pi/\omega$, ... Thus if $\lambda \neq \omega/N$ for each integer N, then the harmonically forced Duffing equation possesses a (unique) family of periodic responses for small $|\beta|$, each with a period of the harmonic input (forcing function), which converges to $x = [F/(\omega^2 - \lambda^2)] \cos \lambda t$ as $\beta \to 0$.

EXAMPLE 3

Consider the Mathieu equation $d^2x/dt^2 + (\omega^2 + \epsilon \cos t)x = 0$. Since it is homogeneous and linear, it is its own equation of first variation. Thus, if the Mathieu equation possesses a periodic solution of period 2π, so also does its equation of first variation. It is not surprising then to find that periodic solutions of period 2π occur only for very special values of the parameters ω^2 and ϵ. These special parameter values map out certain curves of the $\omega^2\epsilon$ plane. If $\omega^2 = f(\epsilon)$ is the equation of a portion of one of these curves, then the Mathieu equation $d^2x/dt^2 + (f(\epsilon) + \epsilon \cos t)x = 0$ possesses a periodic solution of period 2π for each ϵ in some interval. For each such ϵ, the equation of first variation also possesses a periodic solution of period 2π. Thus, the existence of periodic solutions of the equation of first variation of a system does not preclude the possibility of a continuous family of periodic solutions of that system.

EXERCISES

1. In the notation of Theorem 1, show that the solution $\bar{x}(t,\beta,\bar{c})$ is periodic with period τ if and only if it satisfies (4).

2. Show that the Jacobian matrix (6) satisfies (7).

3. Show that (2) possesses a periodic solution of period τ (other than $\bar{y} = \bar{0}$) if and only if the determinant in (9) is zero. First, explain why (2) possesses a nontrivial periodic solution of period τ if and only if there exists a constant vector $\bar{c} \neq \bar{0}$ such that $Y(t + \tau)\bar{c} = Y(t)\,\bar{c}$ for all t, where $Y(t)$ is the principal matrix solution. Then observe that $Y(t + \tau) = Y(t)\,Y(\tau)$ for all t and hence that the above is equivalent to the equation $Y(\tau)\,\bar{c} = \bar{c}$. Complete the argument.

4. Explain why there exists a matrix B such that $Y(\tau) = e^{B\tau}$, where $Y(t)$ is the principal matrix solution of (8). In the present circumstance, the characteristic numbers of B are referred to as characteristic exponents. Show that if zero is a characteristic

exponent, then one is a characteristic multiplier, and hence (2) possesses a periodic solution of period τ. In what sense is there a "converse" to this?

5. Discuss the possible periodic responses of the harmonically forced van der Pol equation

$$\frac{d^2x}{dt^2} + \mu(x^2 - 1)\frac{dx}{dt} + x = F \cos \lambda t$$

for small $|\mu|$.

6. Discuss the possible periodic responses of the harmonically forced Mathieu equation $d^2x/dt^2 + (\omega^2 + \epsilon \cos t)x = F \cos \lambda t$ for small $|\epsilon|$.

7. Let $\bar{c} = \bar{c}(\beta)$ be the continuously differentiable vector function introduced in the proof of Theorem 1. Show that its derivative $d\bar{c}/d\beta$ satisfies the equation

$$\bar{x}_{\bar{c}}\frac{d\bar{c}}{d\beta} + \frac{d\bar{x}}{d\beta} - \frac{d\bar{c}}{d\beta} = \bar{0}$$

where $\bar{x} = \bar{x}(\tau,\beta,\bar{c})$ is the quantity introduced in the proof of Theorem 1, and hence for small $|\beta|$, that $d\bar{c}/d\beta = (I - \bar{x}_{\bar{c}})^{-1}\dfrac{d\bar{x}}{d\beta}$. Thus, since $\bar{c}(0) = \bar{0}$, $\bar{c}^{(1)}(\beta) \equiv (I - \bar{x}_{\bar{c}})_0^{-1}(d\bar{x}/d\beta)_0\beta$, where the subscript "0" denotes the value of the quantities for $\beta = 0$, is a good first approximation to $\bar{c}(\beta)$. One might define a sequence of successive approximations by the recursion formula, $\bar{c}^{(k+1)} = (I - \bar{x}_{\bar{c}})_0^{-1}[\bar{x}(\tau,\beta,\bar{c}^{(k)}) - \bar{p}(0) - (\bar{x}_{\bar{c}})_0\bar{c}^{(k)}]$. Show that this sequence of successive approximations converges to the unique solution $\bar{c} = \bar{c}(\beta)$ of (4). Note that $\bar{x}(\tau,\beta,\bar{c})$ is continuously differentiable in each variable.

8. Explain why the periodic solution $\bar{x}(t,\beta)$ constructed in Theorem 1 is analytic in β if the right-hand member in (1) is analytic in $\bar{x}$ and β. Show how to construct the power-series expansion of $\bar{x}(t,\beta)$ about $\beta = 0$.

◆ ◆ ◆ ◆ ◆ ◆ ◆

It is noteworthy that Theorem 1 is a comparison theorem concerning, in general, nonlinear systems throughout (see Example 4). There is no necessity for (1) to be linear for $\beta = 0$. However, in many situations this is just the crux of the matter. As in Example 2 above, it may be possible to transfer some of the understanding of a linear system to a nonlinear one, at least for small changes. In this connection, one might consider a continuation process, somewhat like the familiar analytical continuation process, as a technique for studying large changes. As the parameter β, say, is increased, Theorem 1 guarantees the existence of a periodic solution $\bar{x}(t,\beta)$ for each β in a neighborhood of a particular value β_0 so long as the corresponding equation of first variation with respect to $\bar{x}(t,\beta_0)$ has no periodic solution of the same period (other than $\bar{y} = \bar{0}$). The process is interrupted only if one encounters a β_0 for which this is not true. The continuation process guarantees that there will be a smallest such β_0. This is not viewed as a practical procedure, however, any more than is analytical continuation.

If the equation of first variation (2) is asymptotically stable at the origin, then certainly it does not admit a (nontrivial) periodic solution of any sort. Thus, Theorem 1 guarantees the existence of periodic solutions for $\beta \neq 0$ which are near the given one $\bar{p}(t)$. We refer to the periodic solution $\bar{p}(t)$ as a *generator* of the family $\bar{x}(t,\beta)$ of periodic solutions. But even more, if (2) is asymptotically stable at the origin, it follows that the corresponding equation of first variation for $\beta \neq 0$, with $|\beta|$ sufficiently small, is also asymptotically stable at the origin. Indeed, this is a direct application of the second part of Theorem 8 of Chap. 4. For, with $|\beta|$ small, the equation of first variation uniformly approximates (2); i.e., $\bar{f}_{\bar{x}}(\bar{x}(t,\beta),\ t,\beta)$ uniformly approximates $\bar{f}_{\bar{x}}(\bar{p}(t),\ t,0)$. Thus we have Theorem 2.

Theorem 2

If the equation of first variation (2) is asymptotically stable at the origin, then for each $|\beta|$ sufficiently small, (1) possesses a (unique) asymptotically stable solution $\bar{x}(t,\beta)$ which is periodic with period τ and for which $\lim_{\beta \to 0} \bar{x}(t,\beta) = \bar{p}(t)$, uniformly for all t.

PROOF: Apply Theorem 1 of the present chapter, Theorem 8 of Chap. 4, and Theorem 3 of Chap. 5, in that order.   The details are left for the student to supply.

EXAMPLE 4

Applications of either Theorem 1 or 2 when the generator $\bar{p}(t)$ is a trivial periodic solution (a singular solution) present interesting variations.   The quasi-linear system

$$\frac{d\bar{x}}{dt} = A\bar{x} + \beta\bar{f}(\bar{x},t) \tag{10}$$

where A is constant and $\bar{f}(\bar{x},t)$ is continuously differentiable and periodic in t with period $\tau > 0$, is an illustration.   If the linear system $d\bar{x}/dt = A\bar{x}$ admits no periodic solutions of period τ (other than $\bar{x} = \overline{0}$), then, for each $|\beta|$ sufficiently small, the quasi-linear system possesses a (unique) periodic solution $\bar{x}(t,\beta)$ of period τ, with $\bar{x}(t,\beta) \to \overline{0}$ as $\beta \to 0$.   If the origin is *not* a singular point of the quasi-linear system, then the periodic solutions for $\beta \neq 0$ are not trivial, even though the generator is.   Of course, each periodic solution is asymptotically stable if the linear system is asymptotically stable at the origin.

EXERCISES

1. Give a detailed proof of Theorem 2.

2. Explain the applications of Theorems 1 and 2 made in Example 4.

3. Explain why the quasi-linear system (10) does not exhibit nontrivial periodic motions in a neighborhood of the origin if the origin is a singular solution for sufficiently small $|\beta|$.

4. Explain why the equation

$$\frac{d^2x}{dt^2} + k\frac{dx}{dt} + \omega^2 x + \beta x^3 = \beta F_0 \cos \lambda t$$

where $k > 0$ and $\omega^2 > 0$, possesses, for each $|\beta|$ sufficiently small, a unique, asymptotically stable periodic solution $x(t,\beta)$ of period

$2\pi/\lambda$ for which $x(t,\beta) \to 0$ as $\beta \to 0$, uniformly for all t.   What is the analogous result if one has $k < 0$?   Note the rather sharp contrast these results make with those derived for an autonomous system in Chap. 6.

5. Consider the equation $d^2x/dt^2 + g(x) = \beta \cos \lambda t$, where the nonlinear spring term $g(x)$ is continuously differentiable, vanishes for $x = 0$, and satisfies $g'(0) \neq 0$.   Discuss the possible periodic solutions of period $2\pi/\lambda$ of this equation for small $|\beta|$.

2. *Periodic Solutions of Nonautonomous Quasi-harmonic Equations*

An important situation, not covered in Theorem 1, is represented by the *quasi-harmonic equation*

$$\frac{d^2x}{dt^2} + x = \beta f\left(x, \frac{dx}{dt}, t, \beta\right) \tag{11}$$

where $f(x, y, t, \beta)$ is continuously differentiable in each variable and periodic in t with period 2π.   In this case, the variational equation for $\beta = 0$ is the equation $d^2y/dt^2 + y = 0$ and so possesses not one but two (independent) periodic solutions of period 2π.   Thus, in the search for periodic solutions of (11) of period 2π, a fresh attack is called for.   We again resort to the integral-equation technique which we have used repeatedly throughout the earlier chapters.

Using (44) of Chap. 4, (see also Example 4 of that chapter) we conclude that any solution of (11) satisfies the integral equation

$$x(t) = a \cos t + b \sin t + \beta \int_0^t \sin (t - s) f\left(x(s), \frac{dx}{ds}, s, \beta\right) ds \tag{12}$$

with

$$\frac{dx}{dt}(t) = -a \sin t + b \cos t + \beta \int_0^t \cos (t - s) f\left(x(s), \frac{dx}{ds}, s, \beta\right) ds \tag{13}$$

The initial conditions are $x = a$ and $dx/dt = b$ for $t = 0$. The solution is periodic with period 2π if $x = a$ and $dx/dt = 0$ also for $t = 2\pi$. For $\beta \neq 0$, these conditions are met in (12) and (13) if and only if

$$\int_0^{2\pi} \sin s\, f\left(x(s), \frac{dx}{ds}, s, \beta\right) ds = 0$$

$$\int_0^{2\pi} \cos s\, f\left(x(s), \frac{dx}{ds}, s, \beta\right) ds = 0 \tag{14}$$

We may regard the general solution of (12) [and hence of (13)] as a function $x(t, \beta, a, b)$ of β and the initial values a and b as well as t. Then the system (14) may be expressed in the form

$$H_1(a,b,\beta) = \int_0^{2\pi} \sin s\, f\left(x(s, \beta, a, b),\right.$$

$$\left.\frac{dx}{ds}(s, \beta, a, b), s, \beta\right) ds = 0$$

$$H_2(a,b,\beta) = \int_0^{2\pi} \cos s\, f\left(x(s, \beta, a, b),\right. \tag{15}$$

$$\left.\frac{dx}{ds}(s, \beta, a, b), s, \beta\right) ds = 0$$

Suppose that this algebraic system possesses the solution $a = a_0$, $b = b_0$ for $\beta = 0$. That is, suppose that

$$H_1(a_0, b_0, 0) = \int_0^{2\pi} \sin s\, f(a_0 \cos s + b_0 \sin s,$$

$$-a_0 \sin s + b_0 \cos s, s, 0)\, ds = 0$$

$$H_2(a_0, b_0, 0) = \int_0^{2\pi} \cos s\, f(a_0 \cos s + b_0 \sin s, \tag{16}$$

$$-a_0 \sin s + b_0 \cos s, s, 0)\, ds = 0$$

We inquire into the possibility of a solution for $\beta \neq 0$ passing to $a = a_c$, $b = b_0$ as $\beta \to 0$. The pertinent Jacobian is given by

$$J(\beta,a,b) = \begin{vmatrix} \dfrac{\partial H_1}{\partial a} & \dfrac{\partial H_1}{\partial b} \\[2mm] \dfrac{\partial H_2}{\partial a} & \dfrac{\partial H_2}{\partial b} \end{vmatrix} \tag{17}$$

and needs to be evaluated for $a = a_0$, $b = b_0$, and $\beta = 0$. We note that

$$\frac{\partial H_1}{\partial a} = \int_0^{2\pi} \sin s \left[f_1 x_a + f_2 \frac{dx_a}{ds} \right] ds$$

$$\frac{\partial H_1}{\partial b} = \int_0^{2\pi} \sin s \left[f_1 x_b + f_2 \frac{dx_b}{ds} \right] ds$$

$$\frac{\partial H_2}{\partial a} = \int_0^{2\pi} \cos s \left[f_1 x_a + f_2 \frac{dx_a}{ds} \right] ds$$

$$\frac{\partial H_2}{\partial b} = \int_0^{2\pi} \cos s \left[f_1 x_b + f_2 \frac{dx_b}{ds} \right] ds$$

where

$$f_1 = \frac{\partial f}{\partial x} \qquad f_2 = \frac{\partial f}{\partial y} \qquad x_a = \frac{\partial x}{\partial a} \qquad x_b = \frac{\partial x}{\partial b}$$

For $\beta = 0$,

$$x = a \cos s + b \sin s \qquad \frac{dx}{ds} = -a \sin s + b \cos s$$

$$x_a = \cos s \qquad \frac{dx_a}{ds} = -\sin s \qquad x_b = \sin s \qquad \text{and} \qquad \frac{dx_b}{ds} = \cos s$$

and if these are evaluated for $a = a_0$, $b = b_0$ and substituted into the above, one obtains for $J(0, a_0, b_0)$ an explicit expression in terms of the solution (a_0, b_0) of (16). If $J(0, a_0, b_0) \neq 0$, then for each $|\beta|$ sufficiently small, the algebraic system (15) possesses a (unique) solution

$$a = a(\beta), b = b(\beta)$$

with $\lim_{\beta \to 0} a(\beta) = a_0$ and $\lim_{\beta \to 0} b(\beta) = b_0$. The corresponding periodic solutions $x(t,\beta) = x(t, \beta, a(\beta), b(\beta))$ of (11) converge (uniformly in t) as $\beta \to 0$ to the generating solution

$$x(t,0) = a_0 \cos t + b_0 \sin t$$

It should be noted that we have discovered the periodic solutions imbedded within a general family of solutions. In this regard, the treatment is similar to that used in the proof of Theorem 1. However, in the present case, one first determines

a candidate for the generating solution using (16), and only then
seeks the neighboring periodic solutions. It is clear that (16)
is a necessary condition for the existence of generator of a family
of periodic solutions of (11) with $|\beta|$ small. In general, the
solution pairs (a_0, b_0) of (16) will be isolated. Those which correspond to actual generators of periodic solutions are referred to
as *bifurcation pairs*. This has reference to the amplitude

$$A(\beta) = [a^2(\beta) + b^2(\beta)]^{1/2}$$

plots of the periodic solutions. The curves $A(\beta)$ versus β
generally extend from bifurcation points $A_0 = [a_0{}^2 + b_0{}^2]^{1/2}$ along

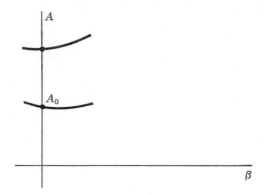

Figure 1

$\beta = 0$ (see Fig. 1). Along $\beta = 0$, every amplitude represents a
periodic solution of (11), while for $\beta \neq 0$, typically only selected
amplitudes represent periodic solutions of (11). The latter stem
from what are called *bifurcation amplitudes*. The question of the
stability of the periodic solutions could be investigated in much
the same manner as the orbital stability of autonomous systems,
(Example 12, Chap. 5) but will not be pursued here.[1] Suffice it
to remark that only orbital stability can in general be anticipated.

[1] For the analytical case, see S. Lefschetz, "Differential Equations:
Geometric Theory," pp. 299–303, Interscience Publishers, Inc., New York,
1957.

EXAMPLE 5

Consider the harmonically forced Duffing equation

$$\frac{d^2x}{dt^2} + (1 + \beta\sigma)x + \beta x^3 = \beta F_0 \cos t$$

Here, σ and $F_0 \neq 0$ are fixed constants. The forcing function is "soft" when $|\beta|$ is small and the "natural" frequencies of the homogeneous systems are nearly unity. In this case we have $f(x,t,\beta) = -\sigma x - x^3 + F_0 \cos t$ for the right-hand member in (11) and (16) becomes

$$H_1(a_0, b_0, 0) = \int_0^{2\pi} [- \sin s(a_0 \cos s + b_0 \sin s)\sigma$$
$$- \sin s(a_0 \cos s + b_0 \sin s)^3 + F_0 \sin s \cos s]\, ds = 0$$
$$H_2(a_0, b_0, 0) = \int_0^{2\pi} [- \cos s(a_0 \cos s + b_0 \sin s)\sigma$$
$$- \cos s(a_0 \cos s + b_0 \sin s)^3 + F_0 \cos^2 s]\, ds = 0$$

When these are expanded and the integrals are evaluated, one finds that $b_0 = 0$ and that a_0 satisfies

$$F_0 - \sigma a_0 - \tfrac{3}{4}a_0^3 = 0 \tag{18}$$

The requirement $J(0, a_0, b_0) = J(0, a_0, 0) \neq 0$, where J is given by (17) reduces merely to the requirement that a_0 be a simple root of (18). Thus, excepting those cases of a double root, (18) yields either one or three bifurcation amplitudes $|a_0|$. [Of course, only real roots of (18) are significant here.] Figure 2 illustrates the various cases. Of course, the double roots occur at the turning points of the constant F_0 curves of Fig. 2.

EXERCISES

1. Explain why (16) is a necessary condition for the existence of a generator of a family of periodic solutions of (11) with $|\beta|$ small.

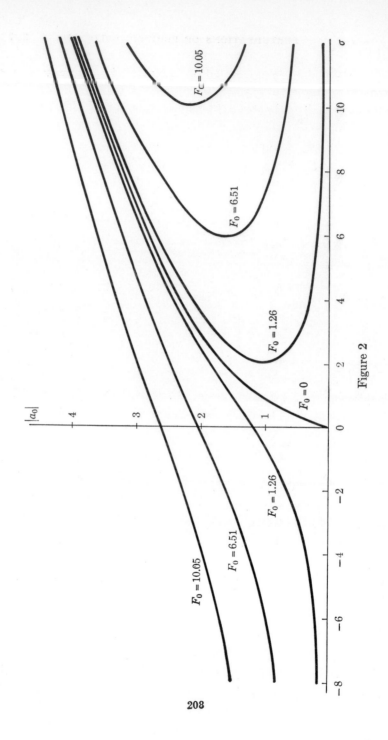

Figure 2

2. Verify the results stated in Example 5. In particular, derive (18) and show that $J = 0$, where J is given by (17), if and only if (18) possesses a double root.

3. Discuss the special significance of the $F_0 = 0$ curve in Fig. 2. Explain how it might be computed *exactly* for each β. Notice that (18) possesses a double root along the $F_0 = 0$ curve. Obtain an equation for the locus along which (18) possesses a double root for $F_0 \neq 0$.

4. Explain how to reformulate the equation

$$\frac{d^2x}{dt^2} + \omega^2 x + \beta x^3 = F \cos \lambda t$$

for small $|F|$ and for $|\lambda|$ nearly equal to $|\omega|$, so as to apply the results of Example 5 in a study of the periodic solutions of period $2\pi/\lambda$. Describe the corresponding amplitude-frequency plot with ω^2 fixed, F a parameter, and frequency λ as independent variable.

5. Discuss the possible periodic solutions of period 2π of the slightly damped, harmonically forced Duffing equation

$$\frac{d^2x}{dt^2} + \beta k \frac{dx}{dt} + (1 + \beta\sigma)x + \beta x^3 = \beta F_0 \cos t$$

Note that the essential role of a small damping term is to link together the separate branches of the constant F_0 curves of Fig. 2.

3. Perturbations of Periodic Solutions in Autonomous Systems

The autonomous case is at once more difficult to treat than is the nonautonomous case, since the equation of first variation, with respect to a periodic solution, always possesses a similar periodic solution. This necessitates some modification not only in the basic analysis, but also in the anticipated results. For if the continuously differentiable, autonomous system

$$\frac{d\bar{x}}{dt} = \bar{f}(\bar{x}, \beta) \tag{19}$$

possesses a nontrivial periodic solution $\bar{p}(t)$ of period τ_0, say, for $\beta = 0$, then the most one can anticipate for (19) with $\beta \neq 0$, is that it possess a nontrivial periodic solution $\bar{x}(t,\beta)$ of period $\tau(\beta)$ (in general, different from τ_0) with $\lim_{\beta \to 0} \tau(\beta) = \tau_0$ and

$$\lim_{\beta \to 0} \bar{x}(t,\beta) = \bar{p}(t)$$

uniformly on each finite interval of t. In favorable circumstances, we shall show that the latter does occur.

If $\bar{p}(t)$ is a nontrivial periodic solution of (19) for $\beta = 0$, then the equation of first variation

$$\frac{d\bar{y}}{dt} = \bar{f}_{\bar{x}}(\bar{p}(t),\, 0)\bar{y} \tag{20}$$

possesses the nontrivial periodic solution $d\bar{p}/dt$. Since (19) is autonomous, we may assume that the origin for t is chosen so that $d\bar{p}/dt \neq \bar{0}$ at $t = 0$. Then by a suitable linear transformation $\bar{u} = B\bar{y}$, $|B| \neq 0$, (20) may be transformed to an equivalent linear system whose principal matrix solution contains the transformed periodic solution $\bar{u}^* = B\, d\bar{p}/dt$ as a column vector. This is accomplished by a rotation (which aligns one of the new coordinate axes with $d\bar{p}/dt$ for $t = 0$) followed by a simple scale factor change (which renders $\bar{u}^*$ a unit vector for $t = 0$). The same linear transformation may be applied to (19) with the result that $B\bar{p}(t)$ becomes the corresponding periodic solution for $\beta = 0$ in the transformed coordinates. Thus, there is no loss in generality if we assume from the outset that in the original coordinate system, $d\bar{p}/dt$ appears as a column, say the first column, of the principal matrix solution $Y(t)$ of (20).

Now denote by $\bar{x}(t,\beta,\bar{c})$, the (general) solution of (19) satisfying the initial condition $\bar{x}(0,\beta,\bar{c}) = \bar{p}(0) + \bar{c}$. Clearly,

$$\bar{x}(t,0,\bar{0}) = \bar{p}(t)$$

As in the proof of Theorem 1, it follows that the Jacobian matrix

$\bar{x}_{\hat{c}}$ is the principal matrix solution of the equation of first variation of (19) with respect to $\bar{x}$. In particular for $\beta = 0$ and $\bar{c} = \bar{0}$, the Jacobian matrix $\bar{x}_{\hat{c}}$ is the principal matrix solution $Y(t)$ of (20). The characteristic multipliers of the periodic system (20) are the roots λ of the algebraic equation

$|Y(\tau_0) - \lambda I| = |\bar{x}_{\hat{c}} - \lambda I|$

$$
= \begin{vmatrix} \dfrac{\partial x_1}{\partial c_1} - \lambda & \dfrac{\partial x_1}{\partial c_2} & \cdots & \dfrac{\partial x_1}{\partial c_n} \\[2mm] \dfrac{\partial x_2}{\partial c_1} & \dfrac{\partial x_2}{\partial c_2} - \lambda & \cdots & \dfrac{\partial x_2}{\partial c_n} \\[2mm] \cdots & \cdots & \cdots & \cdots \\[2mm] \dfrac{\partial x_n}{\partial c_1} & \dfrac{\partial x_n}{\partial c_2} & \cdots & \dfrac{\partial x_n}{\partial c_n} - \lambda \end{vmatrix} = 0 \quad (21)
$$

As indicated above, we may assume that the first column of $Y(t)$ is periodic with period τ_0, and it then follows that the first column of $Y(\tau_0)$ contains the entries $1, 0, \ldots, 0$. Thus, the first column in (21) contains the entries $1 - \lambda, 0, \ldots, 0$, and (21) may be reduced to the product

$$
(1 \quad \lambda) \begin{vmatrix} \dfrac{\partial x_2}{\partial c_2} - \lambda & \dfrac{\partial x_2}{\partial c_3} & \cdots & \dfrac{\partial x_2}{\partial c_n} \\[2mm] \dfrac{\partial x_3}{\partial c_2} & \dfrac{\partial x_3}{\partial c_3} - \lambda & \cdots & \dfrac{\partial x_3}{\partial c_n} \\[2mm] \cdots & \cdots & \cdots & \cdots \\[2mm] \dfrac{\partial x_n}{\partial c_2} & \dfrac{\partial x_n}{\partial c_3} & \cdots & \dfrac{\partial x_n}{\partial c_n} - \lambda \end{vmatrix} = 0 \quad (22)
$$

This equation exhibits, in bold fashion, the fact that $\lambda = 1$ is a characteristic multiplier of the periodic system (20). We consider now the case where $\lambda = 1$ is but a simple root of (22). We have Theorem 3.

THEOREM 3

If $\lambda = 1$ is a simple characteristic multiplier of the equation of first variation (20), then for each $|\beta|$ sufficiently small, there exists a (unique) periodic solution $\bar{x}(t,\beta)$ of period $\tau(\beta)$ of (19) with $\lim_{\beta \to 0} \tau(\beta) = \tau_0$ and $\lim_{\beta \to 0} \bar{x}(t,\beta) = \bar{p}(t)$, uniformly on each finite interval for t.

PROOF: We consider the subfamily of solutions $\bar{x}(t,\beta,\bar{c}^*)$ of (19) for which the initial vector $\bar{p}(0) + \bar{c}^*$ in each case has for its first component, the first component of $\bar{p}(0)$. Thus, the first component of $\bar{c}^*$ is always zero. For $\bar{c}^*$ otherwise arbitrary and β arbitrary, these solutions are not necessarily periodic, but we shall find imbedded within this family of solutions the periodic ones. We restrict considerations throughout to sufficiently small values of $|\beta|$ and $|\bar{c}^*|$ in order that the corresponding solutions exist at least on an internal for t exceeding τ_0 in length, say on $0 \le t \le 2\tau_0$. Then since (19) is autonomous, the solution $\bar{x}(t,\beta,\bar{c}^*)$ will be periodic with period $\tau > 0$ if (and only if)

$$\bar{x}(\tau,\beta,\bar{c}^*) - \bar{p}(0) - \bar{c}^* = \bar{0} \tag{23}$$

Note that for β given, (23) is a system of n equations in the n variables $\tau, c_2, \ldots, c_n$. For $\beta = 0$, the system possesses the solution $\tau = \tau_0$, $\bar{c}^* = \bar{0}$. It will, for sufficiently small $|\beta| \ne 0$, also possess a (unique) solution $[\tau(\beta), \bar{c}^*(\beta)]$ provided the Jacobian, with respect to the variables $\tau, c_2, \ldots, c_n$, does not vanish for $\beta = 0$, $\tau = \tau_0$, $\bar{c}^* = \bar{0}$. But this Jacobian is the determinant

$$\begin{vmatrix} \dfrac{\partial x_1}{dt} & \dfrac{\partial x_1}{\partial c_2} & \dfrac{\partial x_1}{\partial c_3} & \cdots & \dfrac{\partial x_1}{\partial c_n} \\[2ex] \dfrac{\partial x_2}{dt} & \dfrac{\partial x_2}{\partial c_2} - 1 & \dfrac{\partial x_2}{\partial c_3} & \cdots & \dfrac{\partial x_2}{\partial c_n} \\[2ex] \dfrac{\partial x_3}{dt} & \dfrac{\partial x_3}{\partial c_2} & \dfrac{\partial x_3}{\partial c_3} - 1 & \cdots & \dfrac{\partial x_3}{\partial c_n} \\[2ex] \cdots & \cdots & \cdots & \cdots & \cdots \\[2ex] \dfrac{\partial x_n}{dt} & \dfrac{\partial x_n}{\partial c_2} & \dfrac{\partial x_n}{\partial c_3} & \cdots & \dfrac{\partial x_n}{\partial c_n} - 1 \end{vmatrix}$$

which, when evaluated for $\beta = 0$, $t = \tau_0$, $\bar{c}^* = \bar{0}$, becomes

$$
\begin{vmatrix}
\dfrac{dp_1}{dt} & \dfrac{\partial x_1}{\partial c_2} & \dfrac{\partial x_1}{\partial c_3} & \cdots & \dfrac{\partial x_1}{\partial c_n} \\[2mm]
\dfrac{dp_2}{dt} & \dfrac{\partial x_2}{\partial c_2} - 1 & \dfrac{\partial x_2}{\partial c_3} & \cdots & \dfrac{\partial x_2}{\partial c_n} \\[2mm]
\dfrac{dp_3}{dt} & \dfrac{\partial x_3}{\partial c_2} & \dfrac{\partial x_3}{\partial c_3} - 1 & \cdots & \dfrac{\partial x_3}{\partial c_n} \\[2mm]
\cdots & \cdots & \cdots & & \cdots \\[2mm]
\dfrac{dp_n}{dt} & \dfrac{\partial x_n}{\partial c_2} & \dfrac{\partial x_n}{\partial c_3} & \cdots & \dfrac{\partial x_n}{\partial c_n} - 1
\end{vmatrix}
$$

evaluated for $\beta = 0$, $t = \tau_0$, $\bar{c}^* = \bar{0}$. The first column is the periodic solution $d\bar{p}/dt$, which, we have assumed, occupies the first column of the principal matrix solution $Y(t)$. Thus, the pertinent Jacobian vanishes if (and only if)

$$
\begin{vmatrix}
1 & \dfrac{\partial x_1}{\partial c_2} & \dfrac{\partial x_1}{\partial c_3} & \cdots & \dfrac{\partial x_1}{\partial c_n} \\[2mm]
0 & \dfrac{\partial x_2}{\partial c_2} - 1 & \dfrac{\partial x_2}{\partial c_3} & \cdots & \dfrac{\partial x_2}{\partial c_n} \\[2mm]
0 & \dfrac{\partial x_3}{\partial c_2} & \dfrac{\partial x_3}{\partial c_3} - 1 & \cdots & \dfrac{\partial x_3}{\partial c_n} \\[2mm]
\cdots & \cdots & \cdots & & \cdots \\[2mm]
0 & \dfrac{\partial x_n}{\partial c_2} & \dfrac{\partial x_n}{\partial c_3} & \cdots & \dfrac{\partial x_n}{\partial c_n} - 1
\end{vmatrix}
$$

$$
= \begin{vmatrix}
\dfrac{\partial x_2}{\partial c_2} - 1 & \dfrac{\partial x_2}{\partial c_3} & \cdots & \dfrac{\partial x_2}{\partial c_n} \\[2mm]
\dfrac{\partial x_3}{\partial c_2} & \dfrac{\partial x_3}{\partial c_3} - 1 & \cdots & \dfrac{\partial x_3}{\partial c_n} \\[2mm]
\cdots & \cdots & & \cdots \\[2mm]
\dfrac{\partial x_n}{\partial c_2} & \dfrac{\partial x_n}{\partial c_3} & \cdots & \dfrac{\partial x_n}{\partial c_n} - 1
\end{vmatrix} = 0
$$

for $t = \tau_0$ and $\bar{c}^* = \bar{c} = \bar{0}$. Clearly, this is equivalent to the vanishing of the determinant factor in (22) for $\lambda = 1$. But this is impossible since $\lambda = 1$ is a *simple* characteristic multiplier of (20). Hence for each $|\beta|$ sufficiently small, (23) possesses a (unique) solution $[\tau(\beta), \bar{c}^*(\beta)]$ with

$$\lim_{\beta \to 0} \tau(\beta) = \tau_0 \quad \text{and} \quad \lim_{\beta \to 0} \bar{c}^*(\beta) = \overline{0}$$

The corresponding solution $x(t,\beta) = x(t,\beta,\bar{c}^*(\beta))$ of (19) is periodic with period $\tau(\beta)$ and (from the general continuity theorem of Chap. 3) $\lim_{\beta \to 0} \bar{x}(t,\beta) = \lim_{\beta \to 0} x(t,\beta,\bar{c}^*(\beta)) = \bar{p}(t)$, uniformly on each finite interval for t, as stated.

In analogy with the nonautonomous case, it follows immediately that if the number one appears as a simple characteristic multiplier of (20) and if each of the remaining $n - 1$ characteristic multipliers is less than one in absolute value, then for each $|\beta|$ sufficiently small, the periodic solution $\bar{x}(t,\beta) = \bar{x}(t,\beta,\bar{c}^*(\beta))$ is asymptotically orbitally stable.

EXERCISES

1. Referring to characteristic exponents, (in lieu of characteristic multipliers), rephrase Theorem 3 and the result following the proof of Theorem 3.

2. Explain and prove the result following the proof of Theorem 3.

3. Explain in detail the role of the implicit function theorem in the proof of Theorem 3.

4. State and prove a result analogous to Theorem 3 for analytical systems. Explain why the relevant period $\tau(\beta)$ is necessarily analytic in β. Discuss the relationship between this and the Poincaré expansion theorem in Chap. 3.

5. Construct an example illustrating Theorem 3.

4. Periodic Solutions of Autonomous Quasi-harmonic Equations

Theorem 3 is not applicable to the important quasi-harmonic system

$$\frac{d^2x}{dt^2} + x = \beta f\left(x, \frac{dx}{dt}, \beta\right) \tag{24}$$

As in the nonautonomous case, we rephrase (24) as a pair of equivalent integral equations,

$$x(t) = a \cos t + b \sin t + \beta \int_0^t \sin (t - s) f\left(x, \frac{dx}{ds}, \beta\right) ds$$

$$\frac{dx(t)}{dt} = -a \sin t + b \cos t \tag{25}$$

$$+ \beta \int_0^t \cos (t - s) f\left(x, \frac{dx}{ds}, \beta\right) ds$$

However, for $\beta \neq 0$, one cannot anticipate a periodic solution of (24) of period 2π, and so when one imposes the periodicity conditions $x(\tau) = x(0)$, $dx(\tau)/dt = dx(0)/dt$ for a period $\tau = 2\pi + \eta$ the resulting equations no longer contain just the integral parts. We obtain, instead, the equations

$$H(\eta,a,b) = a(\cos \eta - 1) + b \sin \eta$$

$$+ \beta \int_0^{2\pi+\eta} \sin (\eta - s) f\left(x, \frac{dx}{ds}, \beta\right) ds = 0$$

$$K(\eta,a,b) = -a \sin \eta + b(\cos \eta - 1) \tag{26}$$

$$+ \beta \int_0^{2\pi+\eta} \cos (\eta - s) f\left(x, \frac{dx}{ds}, \beta\right) ds = 0$$

As they stand, these equations are degenerate for $\beta = 0$; i.e., Eqs. (26) are satisfied identically in a and b for $\eta = 0$ or identically in η for $a = b = 0$. In the forced case, wherein only the integral parts appeared, this degeneracy was removed merely upon division by $\beta \neq 0$. This step, though trivial in that case, was essential to the argument.   We shall again be able to remove the degeneracy here in an analogous, but less trivial, fashion.

It will be sufficient to exhibit solutions of (26) in which $a = 0$ and $\eta = \beta \eta_1$.   Thus we consider the system

$$
\begin{aligned}
H(\beta \eta_1, 0, b) = H(\eta_1, b) &= b \sin \beta \eta_1 \\
&+ \beta \int_0^{2\pi + \beta \eta_1} \sin (\beta \eta_1 - s) f\left(x, \frac{dx}{ds}, \beta\right) ds = 0 \\
K(\beta \eta_1, 0, b) = K(\eta_1, b) &= b(\cos \beta \eta_1 - 1) \\
&+ \beta \int_0^{2\pi + \beta \eta_1} \cos (\beta \eta_1 - s) f\left(x, \frac{dx}{ds}, \beta\right) da = 0
\end{aligned}
\tag{27}
$$

Now upon division by $\beta \neq 0$, there results the equivalent system

$$
\begin{aligned}
b \eta_1 + \beta^2 L_1 + \int_0^{2\pi + \beta \eta_1} \sin (\beta \eta_1 - s) f\left(x, \frac{dx}{ds}, \beta\right) ds &= 0 \\
-\frac{b}{2} \beta \eta_1{}^2 + \beta^3 L_2 + \int_0^{2\pi + \beta \eta_1} \cos (\beta \eta_1 - s) f\left(x, \frac{dx}{ds}, \beta\right) ds &= 0
\end{aligned}
\tag{28}
$$

where L_1 and L_2 denote terms of no consequence in the following. For $\beta = 0$, the second equation in (28) imposes a requirement upon the corresponding values of $b = b_0$.   Thus, any bifurcation amplitude $|b_0|$ must necessarily satisfy the equation

$$
\Phi(b_0) = \int_0^{2\pi} \cos s \, f(b_0 \sin s, b_0 \cos s, 0) \, ds = 0
\tag{29}
$$

If in addition $b_0 \neq 0$, then the first equation of (28) yields, for $\beta = 0$,

$$
\eta_1 = \psi(b_0) = \frac{1}{b_0} \int_0^{2\pi} \sin s \, f(b_0 \sin s, b_0 \cos s, 0) \, ds
\tag{30}
$$

Thus for $\beta = 0$, the system (28) possesses the solution $b = b_0$, $\eta_1 = \psi(b_0)$, provided b_0 is a nonzero real root of (29).   Further, if the Jacobian, with respect to the pair (b, η_1), of the left-hand members in (28) does not vanish for $\beta = 0$, $b = b_0$, $\eta_1 = \psi(b_0)$, then for each $|\beta|$ sufficiently small there exists a (unique) continuous solution $[b(\beta), \eta_1(\beta)]$ of (28) satisfying $\lim_{\beta \to 0} b(\beta) = b_0$ and $\lim_{\beta \to 0} \eta_1(\beta) = \psi(b_0)$.   The corresponding solution of (24) is periodic with period $\tau = 2\pi + \beta\eta_1(\beta)$ and amplitude $|b(\beta)|$. But with $\beta = 0$, the left-hand members in (28) reduce to the pair.

$$b\eta_1 - \int_0^{2\pi} \sin s \, f(b \sin s, \, b \cos s, \, 0) \, ds$$

and $\qquad \displaystyle\int_0^{2\pi} \cos s \, f(b \sin s, \, b \cos s, \, 0) \, ds$

so that the Jacobian in question is merely the determinant

$$J = \begin{vmatrix} \eta_1 - \psi(b_0) - b_0\psi'(b_0) & b_0 \\ \Phi'(b_0) & 0 \end{vmatrix} = -b_0\Phi'(b_0)$$

Thus if $b_0 \neq 0$ is a *simple* (real) root of (29), then $|b_0|$ is necessarily a bifurcation amplitude for a continuous family of periodic

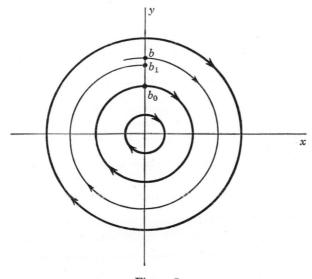

Figure 3

solutions of (24). Each bifurcation amplitude $|b_0|$ is a root (whether simple or not) of (29). Note that if b_0 is a root of (29), so also is $-b_0$. Let us suppose further that (29) possesses only isolated roots. Then in the phase plane, the periodic solutions (periodic orbits) for small $|\beta|$ form a family of concentric ovals enclosing the origin (see Fig. 3). Each oval approximates the

generating solution corresponding to a bifurcation amplitude
$|b_0|$. The latter, of course, is the circle $x^2 + y^2 = b_0^2$ and is
called a *generating circle*. The stability (orbital stability) of the
periodic orbit is reflected in the sign of the function

$$\Phi(b) = \int_0^{2\pi} \cos s \, f(b \sin s, b \cos s, 0) \, ds$$

for b lying between the bifurcation amplitudes. In fact, if a
solution $x(t)$ of (24) crosses the positive y axis for $y = b$, then
the next crossing, after a time lapse equal to $2\pi + \beta\eta_1$, where
η_1 satisfies the first equation in (27), will occur at a point $y = b_1$,
where $b_1 - b = K(\eta_1, b)$ is the left-hand member of the second
equation in (27). Except for the factor β, the latter is also the
left-hand member of the second equation in (28). Hence

$$b_1 - b = -\frac{b}{2}\beta^2\eta_1{}^2 + \beta^4 L_2 + \beta \int_0^{2\pi + \beta\eta_1} \cos(\beta\eta_1 - s) f\left(x, \frac{dx}{ds}, \beta\right) ds \tag{31}$$

For $|\beta|$ sufficiently small, the sign of the right-hand member of
(31) will be that of $\beta\Phi(b)$. The change in the y intercept then
has the same sign as $\beta\Phi(b)$. If $\beta\Phi(b)$ is negative in a semineighborhood $b > b_0$ of a bifurcation amplitude $b_0 > 0$, then the
neighboring exterior trajectories spiral toward the corresponding
periodic orbit. If $\beta\Phi(b)$ is positive in a semineighborhood $b > b_0$,
then the neighboring exterior trajectories spiral away from the
periodic orbit. Analogous statements may be made for the
neighboring interior trajectories. In particular, if $b_0 > 0$ is a
bifurcation amplitude and $\beta\Phi'(b_0) < 0$, then the corresponding
periodic orbit is asymptotically orbitally stable. If $\beta\Phi'(b_0) > 0$,
then the periodic orbit is unstable. The technique employed
here is called the *Poincaré method of sections*. We have, in effect,
reduced the investigation to geometric questions relating to the
successive points of intersection of a phase-plane trajectory with
the positive y axis. The latter is called a *manifold of sections*.
A periodic solution emerges as a fixed point of this manifold

while its orbital stability characteristics are reflected in the "motion" of neighboring points of the manifold.

EXERCISES

1. Apply the method of this section to the van der Pol equation, $d^2x/dt^2 + \mu(x^2 - 1)dx/dt + x = 0$, with $|\mu|$ small. In particular, determine the bifurcation amplitudes and their associated stability characteristics.

2. State and prove results, analogous to those obtained in this section, for analytical systems. Explain why the relevant period $2\pi + \eta(\beta)$ is analytic in β. Discuss the relationship between this and the Poincaré expansion theorem in Chap. 3. In particular, discuss the relevance of Example 3 (and related exercises) of Chap. 3.

3. In the notation of this section, explain why the solution variable η_1 of (27) is given approximately by $\eta_1 = \psi(b)$, for $|\beta|$ small. Use this to discuss the periods of the periodic solutions of (24) and the relative phases of neighboring solutions. Explain the significance of the roots of $\psi(b) = 0$.

4. In the notation of this section, prove that a periodic orbit, corresponding to a bifurcation amplitude $b_0 > 0$, is asymptotically orbitally stable if $\beta\Phi'(b_0) < 0$.

Chapter 8

A GENERAL ASYMPTOTIC METHOD

1. *Introduction*

The theorems of the previous two chapters have focused attention on periodic solutions. This is altogether appropriate since in dynamical systems periodic motion is certainly the first order of business. Still, the theorems offer scarcely a hint as to the nature of the periodic motion and provide essentially nothing at all for practical applications. It is small comfort to an engineer, for example, to know that the steady-state response, whatever its nature, is imbedded within a general family of responses, none of which are available to him. However, fortified by the assurance that a periodic solution exists, it is often possible to uncover some of its more salient properties. Very often, a few terms of its Fourier expansion may be identified and this may be sufficient for many purposes. Some of the "theoretical" results of the previous chapters may indicate the stability characteristics, although more often than not, for rigorous conclusions one will be forced to look elsewhere. A theory of aperiodic solutions is practically nonexistent.

In this final chapter, therefore, we shall illustrate a general perturbational method for the study of linear and nonlinear oscillations. The method satisfies the demand in applied mathematics for a sound and versatile technique and at the same time the requirement of basic simplicity for practical applications. These features are illustrated in a study of the three most poignant and historically prominent examples from the theory of oscillations.

The procedure is to develop, essentially by an iterative process, the solution in the form of one or more asymptotic series in a perturbation parameter. This is merely a formal procedure and nothing is claimed, in regard to convergence, for the asymptotic series. However, the essential properties of a solution for values of the perturbation parameter with small magnitude are reflected in the asymptotic form, and in most applications the asymptotic series itself may be regarded as a solution. There are instances, of course, wherein the asymptotic series actually converges and hence is an exact perturbational solution.

2. *The Mathieu Equation*

Consider the Mathieu equation

$$\frac{d^2u}{dt^2} + \omega^2 u = \epsilon u \cos t \tag{1}$$

For $\epsilon = 0$, the solutions of (1) are the uniformly bounded, simple harmonic functions of period $2\pi/\omega$. For $\epsilon \neq 0$, one would expect the solutions to be somewhat similar. To what extent this expectation is fulfilled is the subject of the present investigation. The periodic solutions of (1) of period 2π or 4π are called *Mathieu functions* and completely characterize the stability properties of (1). However, only for very special values of ω^2 and ϵ do such periodic solutions exist. In general, the solutions of (1) are either almost periodic, or diverging or decaying oscillations. We shall express a general perturbational solution of (1) in the form of an asymptotic series

$$u = A \cos(\omega t - \theta) + \epsilon u_1 + \epsilon^2 u_2 + \epsilon^3 u_3 + \cdots + \epsilon^N u_N \tag{2}$$

where each of A, θ, u_1, u_2, ..., u_N is, in general, a variable. The first term of the expansion represents the principal part of the solution and, with A and θ each variable, suggests a variation of parameters treatment. On the other hand, the additive terms in powers of the parameter ϵ suggest, and provide for, an expansion-type perturbational treatment. The following procedure incorporates what are probably the principal advantages of the two classical techniques and, at the same time, avoids what are probably the principal shortcomings of each.

If (2) is substituted in (1), there results

$$
\begin{aligned}
\left[2\omega A \frac{d\theta}{dt} + \frac{d^2A}{dt^2} - A\left(\frac{d\theta}{dt}\right)^2 \right] &\cos(\omega t - \theta) \\
+ \left[-2\omega \frac{dA}{dt} + 2\frac{dA}{dt}\frac{d\theta}{dt} + A\frac{d^2\theta}{dt^2} \right] &\sin(\omega t - \theta) \\
+ \epsilon\left(\frac{d^2u_1}{dt^2} + \omega^2 u_1\right) + \epsilon^2\left(\frac{d^2u_2}{dt^2} + \omega^2 u_2\right) &+ \cdots \\
+ \epsilon^N\left(\frac{d^2u_N}{dt^2} + \omega^2 u_N\right) &= \frac{\epsilon A}{2}\cos[(\omega+1)t - \theta] \\
+ \frac{\epsilon A}{2}\cos[(\omega-1)t - \theta] + \epsilon^2 u_1 \cos t + \epsilon^3 u_2 \cos t &+ \cdots \\
&+ \epsilon^{N+1}u_N \cos t \quad (3)
\end{aligned}
$$

For $\epsilon = 0$, (3) implies that each of θ and A is a constant, which, of course, is as it should be. The explicit first-order terms (in ϵ) in (3) suggest that u_1 should satisfy the equation

$$
\frac{d^2u_1}{dt^2} + \omega^2 u_1 = \frac{A}{2}\cos[(\omega+1)t - \theta] + \frac{A}{2}\cos[(\omega-1)t - \theta] \quad (4)
$$

Indeed, this is appropriate provided the frequencies on the right do not produce resonance or near resonance in u_1. We may assume that ω is positive throughout and so only the second term on the right in (4) can produce resonance or near resonance. For $\omega = \frac{1}{2}$ and θ constant, this term *is* resonant and every solution of (4) is unbounded. It is therefore necessary to remove the term. It may be interpreted as a fundamental harmonic term in (3).   In fact, this term leads to two harmonics (with $\omega = \frac{1}{2}$),

$$\frac{A}{2}\cos\left[-\frac{t}{2}-\theta\right] = \frac{A}{2}\cos[\tfrac{1}{2}t - \theta + 2\theta]$$

$$= \frac{A}{2}\cos 2\theta \cos[\tfrac{1}{2}t - \theta]$$

$$\frac{A}{2}\sin 2\theta \sin[\tfrac{1}{2}t - \theta]$$

Thus, for $\omega = \frac{1}{2}$ we obtain from the fundamental harmonics in (3),

$$\frac{d\theta}{dt} + \frac{1}{A}\frac{d^2A}{dt^2} - \left(\frac{d\theta}{dt}\right)^2 = \frac{\epsilon}{2}\cos 2\theta$$

$$-\frac{1}{A}\frac{dA}{dt} + \frac{2}{A}\frac{dA}{dt}\frac{d\theta}{dt} + \frac{d^2\theta}{dt^2} = \frac{-\epsilon}{2}\sin 2\theta \tag{5}$$

and (4) is replaced by

$$\frac{d^2u_1}{dt^2} + \tfrac{1}{4}u_1 = \frac{A}{2}\cos(\tfrac{3}{2}t - 0) \tag{6}$$

Fortunately, it is not necessary to treat the variational equations (5) precisely, but rather, it is sufficient at this point to obtain solutions correct to *first order* in ϵ.   Hence we reduce (5) to the system

$$\frac{d\theta}{dt} = \frac{\epsilon}{2}\cos 2\theta \tag{7}$$

$$\frac{1}{A}\frac{dA}{dt} = \frac{\epsilon}{2}\sin 2\theta \tag{8}$$

merely by dropping the terms d^2A/dt^2, $(d\theta/dt)^2$, $(dA/dt\,d\theta/dt)$, and $d^2\theta/dt^2$.   According to (7) and (8), each of the omitted terms will be of second order in ϵ.   From the first equation, (7), we see that $t \to \infty$ as $\cos 2\theta \to 0$.   Using (7), (8) may be expressed in the differential form

$$\frac{dA}{A} = \frac{\sin 2\theta}{\cos 2\theta}\,d\theta = \mp\tfrac{1}{2}d(\ln \cos |2\theta|)$$

Hence if $\cos 2\theta > 0$ initially, then

$$\frac{A}{A_0} = \sqrt{\frac{\cos 2\theta_0}{\cos 2\theta}} \tag{9}$$

and if $\cos 2\theta < 0$ initially, then

$$\frac{A}{A_0} = \sqrt{\frac{\cos 2\theta}{\cos 2\theta_0}} \tag{10}$$

where A_0 and θ_0 may be regarded as initial values of A and θ respectively. In the former case, $|A| \to \infty$ as $t \to \infty$, while in the latter case, $A \to 0$ as $t \to \infty$. Thus for $\omega = \frac{1}{2}$, we have solutions which are diverging oscillations and solutions which are decaying oscillations. If $\cos 2\theta_0 = 0$, then $\theta \equiv \theta_0$ and $A = A_0 e^{\pm \epsilon t/2}$, with the sign being that of $\sin 2\theta_0$.

On the other hand, for ω appreciably different from $\frac{1}{2}$, the solution to first order in ϵ becomes

$$u = A \cos (\omega t - \theta) - \frac{\epsilon A}{2(2\omega + 1)} \cos [(\omega + 1)t - \theta]$$

$$+ \frac{\epsilon A}{2(2\omega - 1)} \cos [(\omega - 1)t - \theta] \tag{11}$$

where each of A and θ is a constant. It is of interest to consider how (11) degenerates as $\omega \to \frac{1}{2}$ and how the diverging and decaying oscillations emerge.

For ω near $\frac{1}{2}$, we may write the troublesome term in (4) in the form

$$\frac{A}{2} \cos [(\omega - 1)t - \theta] = \frac{A}{2} \cos [(1 - 2\omega)t + 2\theta] \cos (\omega t - \theta)$$

$$- \frac{A}{2} \sin [(1 - 2\omega)t + 2\theta] \sin (\omega t - \theta)$$

and, as in the exact resonance case, shift these terms to the fundamental harmonic terms in (3). If we assume that $(1 - 2\omega)$ is small of order ϵ, we may again reduce the full variational equations and obtain the system

$$\frac{d\theta}{dt} = \frac{\epsilon}{4\omega} \cos [(1 - 2\omega)t + 2\theta]$$

$$\frac{1}{A}\frac{dA}{dt} = \frac{\epsilon}{4\omega} \sin [(1 - 2\omega)t + 2\theta]$$

We introduce the auxiliary variable

$$\Phi = (1 - 2\omega)t + 2\theta \tag{12}$$

and these become

$$\frac{d\Phi}{dt} = 1 - 2\omega + \frac{\epsilon}{2\omega} \cos \Phi \tag{13}$$

$$\frac{1}{A}\frac{dA}{dt} = \frac{\epsilon}{4\omega} \sin \Phi \tag{14}$$

The behavior of a solution of this system depends upon the relative sizes of the quantities $|1 - 2\omega|$ and $|\epsilon|/2\omega$. For $|1 - 2\omega| > |\epsilon|/2\omega$, $d\Phi/dt = 1 - 2\omega + (\epsilon/2\omega) \cos \Phi$ never vanishes and Φ is monotone with t, while for $|1 - 2\omega| < |\epsilon|/2\omega$, $d\Phi/dt$ necessarily vanishes. In the former case, all solutions of (14) are bounded, while in the latter case, there are always unbounded solutions. In fact, when (13) is used, (14) may be expressed in the differential form

$$\frac{dA}{A} = \frac{\epsilon/4\omega \sin \Phi \, d\Phi}{1 - 2\omega + \epsilon/2\omega \cos \Phi} = \mp \frac{1}{2} d \left(\ln \left| 1 - 2\omega + \frac{\epsilon}{2\omega} \cos \Phi \right| \right)$$

and so either

$$\frac{A}{A_0} = \sqrt{\frac{1 - 2\omega + \epsilon/2\omega \cos \Phi_0}{1 - 2\omega + \epsilon/2\omega \cos \Phi}}$$

or

$$\frac{A}{A_0} = \sqrt{\frac{1 - 2\omega + \epsilon/2\omega \cos \Phi}{1 - 2\omega + \epsilon/2\omega \cos \Phi_0}}$$

depending upon the initial sign of the quantity $1 - 2\omega + \epsilon/2\omega \cos \Phi$. The locus

$$|1 - 2\omega| = \frac{|\epsilon|}{2\omega} \tag{15}$$

therefore, separates these radically different behaviors. In the $\epsilon\omega$ plane, (15) depicts certain curves, to one side of which the parameter pairs (ϵ,ω) always correspond to bounded solutions of (1) and to the other side of which the parameter pairs (ϵ,ω) correspond to both bounded and unbounded solutions of (1). These curves are illustrated in Fig. 1. It is customary to plot $|\epsilon|$ versus ω^2 rather than ϵ versus ω, and we follow this custom. It is just along these curves that the periodic solutions, cor-

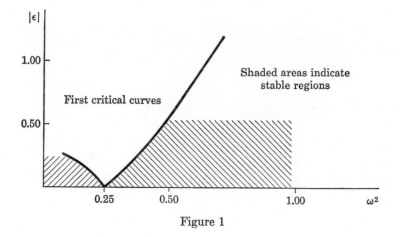

Figure 1

responding to the Mathieu functions, emerge. Typically, the curves themselves are found by ascribing to them this property, i.e., the property of admitting periodic (with period 2π or 4π) solutions of (1). Here we note that Eqs. (13) and (14) become

$$\frac{d\Phi}{dt} = \frac{\epsilon}{2\omega}[\pm 1 + \cos\Phi] \qquad (16)$$

and

$$\frac{1}{A}\frac{dA}{dt} = \frac{\epsilon}{4\omega}\sin\Phi \qquad (17)$$

The periodic solutions correspond to one of the trivial (singular) solutions of the system (16), (17), say $\Phi \equiv \pi$, with the plus sign in (16), or $\Phi \equiv 0$, with the minus sign in (16). In such a case, A

is constant, and from (12) we obtain

$$\frac{d\Phi}{dt} = (1 - 2\omega) + 2\frac{d\theta}{dt} \equiv 0$$

The frequency of the fundamental harmonic in (2), therefore, is given by

$$\omega - \frac{d\theta}{dt} = \omega + \frac{1 - 2\omega}{2} = \frac{1}{2}$$

and the least period is 4π. Of course, there are other (nontrivial) solutions of the system (16), (17), corresponding to other initial values of Φ. For these, (17) may be written

$$\frac{dA}{A} = \frac{1}{2}\frac{\sin \Phi}{\pm 1 + \cos \Phi} d\Phi = \mp \frac{1}{2} d(\ln |\pm 1 + \cos \Phi|)$$

and so $\quad \dfrac{A}{A_0} = \sqrt{\dfrac{\pm 1 + \cos \Phi_0}{\pm 1 + \cos \Phi}} \qquad$ for $\pm 1 + \cos \Phi_0 > 0$

or $\quad \dfrac{A}{A_0} = \sqrt{\dfrac{\pm 1 + \cos \Phi}{\pm 1 + \cos \Phi_0}} \qquad$ for $\pm 1 + \cos \Phi_0 < 0$

Thus, for $|1 - 2\omega| = |\epsilon|/2\omega$, there are bounded and unbounded nonperiodic solutions in addition to the periodic ones.

Higher-order solutions may be obtained for both the resonant and nonresonant cases. We shall consider only the latter here. Thus we now assume that the parameter pair (ϵ, ω) is appreciably away from the critical curves defined by (15). As we shall encounter other critical curves, the curves defined by (15) will be referred to as the *first set of critical curves*. If the first-order solution (11) is substituted in (3), we obtain the second-order variational equations (reduced from the full variational equations)

$$\begin{aligned}\frac{d\theta}{dt} &= \frac{\epsilon^2}{4\omega(4\omega^2 - 1)}\\ \frac{1}{A}\frac{dA}{dt} &= 0\end{aligned} \tag{18}$$

and the second-order perturbational equation

$$\frac{d^2u_2}{dt^2} + \omega^2 u_2 = -\frac{A}{4(2\omega + 1)} \cos\left[(\omega + 2)t - \theta\right]$$

$$+ \frac{A}{4(2\omega - 1)} \cos\left[(\omega - 2)t - \theta\right] \quad (19)$$

Again, if resonance is not a problem in (19), we have immediately the second-order (almost periodic) solution

$$u = A \cos(\omega t - \theta) - \frac{\epsilon A}{2(2\omega + 1)} \cos\left[(\omega + 1)t - \theta\right]$$

$$+ \frac{\epsilon A}{2(2\omega - 1)} \cos\left[(\omega - 1)t - \theta\right]$$

$$+ \frac{\epsilon^2 A}{16(2\omega + 1)(\omega + 1)} \cos\left[(\omega + 2)t - \theta\right]$$

$$+ \frac{\epsilon^2 A}{16(2\omega - 1)(\omega - 1)} \cos\left[(\omega - 2)t - \theta\right] \quad (20)$$

where

$$\theta = \theta_0 + \frac{\epsilon^2 t}{4\omega(4\omega^2 - 1)}$$

and A is constant.   For ω near 1, however, the last term on the right in (20) degenerates, and we must remove the offending term in (19).   We write

$$\frac{A}{4(2\omega - 1)} \cos\left[(\omega - 2)t - \theta\right]$$

$$= \frac{A}{4(2\omega - 1)} \cos\left[2\theta + 2(1 - \omega)t\right] \cos(\omega t - \theta)$$

$$- \frac{A}{4(2\omega - 1)} \sin\left[2\theta + 2(1 - \omega)t\right] \sin(\omega t - \theta)$$

and introduce these in (3) as fundamental harmonics.   Then the system (18) is modified according to

$$\frac{d\theta}{dt} = \frac{\epsilon^2}{4\omega(4\omega^2 - 1)} + \frac{\epsilon^2}{8\omega(2\omega - 1)} \cos\left[2\theta + 2(1 - \omega)t\right]$$

$$\frac{dA}{dt} = \frac{\epsilon^2}{8\omega(4\omega^2 - 1)} \sin\left[2\theta + 2(1 - \omega)t\right]$$

In terms of the auxiliary variable $\Phi = 2\theta + 2(1 - \omega)t$, these become

$$\frac{d\Phi}{dt} = \frac{\epsilon^2}{2\omega(4\omega^2 - 1)} + 2(1 - \omega) + \frac{\epsilon^2}{4\omega(2\omega - 1)} \cos \Phi \quad (21)$$

and $\quad \dfrac{dA}{dt} = \dfrac{\epsilon^2}{8\omega(4\omega^2 - 1)} \sin \Phi \qquad\qquad\qquad (22)$

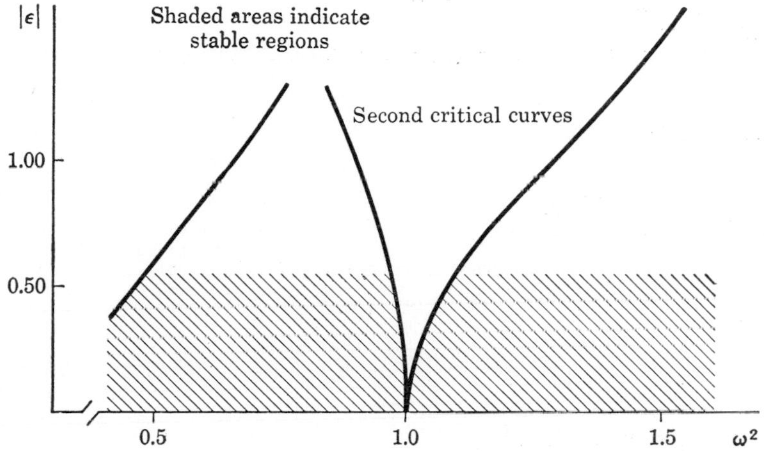

Figure 2

Thus we encounter a second set of critical curves near $\omega = 1$. These are determined by the equation

$$\left| \frac{\epsilon^2}{2\omega(4\omega^2 - 1)} + 2(1 - \omega) \right| = \frac{\epsilon^2}{4\omega(2\omega - 1)} \qquad (23)$$

and are referred to as the *second set of critical curves* (see Fig. 2). For

$$\left| \frac{\epsilon^2}{2\omega(4\omega^2 - 1)} + 2(1 - \omega) \right| \neq \frac{\epsilon^2}{4\omega(2\omega - 1)}$$

the solutions of (22) are given as either

$$\frac{A}{A_0} = \sqrt{\frac{a_1 + b_1 \cos \Phi_0}{a_1 + b_1 \cos \Phi}} \tag{24}$$

or

$$\frac{A}{A_0} = \sqrt{\frac{a_1 + b_1 \cos \Phi}{a_1 + b_1 \cos \Phi_0}} \tag{25}$$

with

$$\tan \frac{\Phi}{2} = \frac{\tan \Phi_0/2 \sqrt{a_1^2 - b_1^2} + (a_1 + b_1) \tan (\sqrt{a_1^2 - b_1^2}\, t/2)}{\sqrt{a_1^2 - b_1^2} - (a_1 - b_1) \tan \Phi_0/2 \tan (\sqrt{a_1^2 - b_1^2}\, t/2)} \tag{26}$$

for $|a_1| > |b_1|$, or

$$\tan \frac{\Phi}{2} = \frac{a_1 + b_1}{\sqrt{b_1^2 - a_1^2}}$$
$$\left\{ \frac{(\sqrt{b_1^2 - a_1^2} \tan \Phi_0/2 + a_1 + b_1) \exp (\sqrt{b_1^2 - a_1^2}\, t) + \sqrt{b_1^2 - a_1^2} \tan \Phi_0/2 - a_1 - b_1}{(\sqrt{b_1^2 - a_1^2} \tan \Phi_0/2 + a_1 + b_1) \exp (\sqrt{b_1^2 - a_1^2}\, t) - \sqrt{b_1^2 - a_1^2} \tan \Phi_0/2 + a_1 + b_1} \right\} \tag{27}$$

for $|a_1| < |b_1|$, where

$$a_1 = \frac{\epsilon^2}{2\omega(4\omega^2 - 1)} + 2(1 - \omega)$$

and

$$b_1 = \frac{\epsilon^2}{4\omega(2\omega - 1)}$$

Along the critical curves (23), periodic solutions correspond to the trivial (singular) solutions of the system

$$\frac{d\Phi}{dt} = \frac{\epsilon^2}{4\omega(2\omega - 1)} (\pm 1 + \cos \Phi)$$
$$\frac{dA}{dt} = \frac{\epsilon^2}{4\omega(2\omega - 1)} \sin \Phi$$

and there are other nonperiodic bounded and unbounded solutions given by (24) and (25). The least period of the periodic solutions in this case is 2π since $0 \equiv d\Phi/dt = 2\,d\theta/dt + 2(1 - \omega)$, and hence $\omega - d\theta/dt = 1$. Thus these periodic solutions also correspond to Mathieu functions.

Considering only the nonresonant case (20), the third-order variational equations become

$$\frac{d\theta}{dt} = \frac{\epsilon^2}{4\omega(4\omega^2 - 1)}$$
$$\frac{dA}{dt} = 0 \tag{28}$$

and the third-order perturbational equation becomes

$$\begin{aligned}
\frac{d^2u_3}{dt^2} + \omega^2 u_3 = {} & \frac{(\omega + 1)A}{4\omega(2\omega + 1)(4\omega^2 - 1)} \cos\left[(\omega + 1)t - \theta\right] \\
& - \frac{(\omega - 1)A}{4\omega(2\omega - 1)(4\omega^2 - 1)} \cos\left[(\omega - 1)t - \theta\right] \\
& + \frac{A}{32(2\omega + 1)(\omega + 1)} \cos\left[(\omega + 3)t - \theta\right] \\
& + \frac{A}{32(2\omega + 1)(\omega + 1)} \cos\left[(\omega + 1)t - \theta\right] \\
& + \frac{A}{32(2\omega - 1)(\omega - 1)} \cos\left[(\omega - 1)t - \theta\right] \\
& + \frac{A}{32(2\omega - 1)(\omega - 1)} \cos\left[(\omega - 3)t - \theta\right] \tag{29}
\end{aligned}$$

It should be pointed out here that (29) incorporates terms (the first two on the right) which correct, to order ϵ^3, the term $\epsilon(d^2u_1/dt^2 + \omega^2 u_1)$ appearing on the left in (3). For u_1 does not satisfy (4) to order ϵ^3 unless $d\theta/dt$ is zero to order ϵ^2. But with ω appreciably different from 1, $d\theta/dt$ is given by (18) to order ϵ^2 and is not zero. If ω is also appreciably different from $\frac{3}{2}$, no resonance occurs in (29) and the third-order solution is given by

$$u = A \cos(\omega t - \theta) + \epsilon u_1 + \epsilon^2 u_2 + \epsilon^3 u_3$$

with

$$u_3 = \frac{A}{32(2\omega - 1)(\omega - 1)(2\omega - 1)} \cos[(\omega - 1)t - \theta]$$

$$- \frac{A}{32(2\omega + 1)^2(\omega + 1)} \cos[(\omega + 1)t - \theta]$$

$$- \frac{A}{96(2\omega + 1)(\omega + 1)(2\omega + 3)} \cos[(\omega + 3)t - \theta]$$

$$- \frac{A}{4\omega(2\omega - 1)^2(4\omega^2 - 1)} \cos[(\omega - 1)t - \theta]$$

$$- \frac{(\omega + 1)A}{4\omega(2\omega + 1)^2(4\omega^2 - 1)} \cos[(\omega + 1)t - \theta]$$

$$+ \frac{A}{96(2\omega - 1)(\omega - 1)(2\omega - 3)} \cos[(\omega - 3)t - \theta] \quad (30)$$

where
$$\theta = \theta_0 + \frac{\epsilon^2 t}{4\omega(4\omega^2 - 1)}$$

and A is constant. However, we encounter a third set of critical curves near $\omega = \frac{3}{2}$, along which the last term on the right in (30) degenerates. The offending term (last term on the right) in (29) is removed and expressed in terms of the fundamentals as before. Equations (28) are then modified according to

$$\frac{d\theta}{dt} = \frac{\epsilon^2}{4\omega(4\omega^2 - 1)} + \frac{\epsilon^3}{64\omega(2\omega - 1)(\omega - 1)} \cos[2\theta + (3 - 2\omega)t]$$

$$\frac{1}{A}\frac{dA}{dt} = \frac{\epsilon^3}{64\omega(2\omega - 1)(\omega - 1)} \sin[2\theta + (3 - 2\omega)t]$$

With $\Phi = 2\theta + (3 - 2\omega)t$, these become

$$\frac{d\Phi}{dt} = \frac{\epsilon^2}{2\omega(4\omega^2 - 1)} + (3 - 2\omega) + \frac{\epsilon^3}{32\omega(2\omega - 1)(\omega - 1)} \cos\Phi$$

$$\frac{1}{A}\frac{dA}{dt} = \frac{\epsilon^3}{64(2\omega - 1)(\omega - 1)} \sin\Phi \qquad (31)$$

The third set of critical curves are defined by the equation

$$\left| \frac{\epsilon^2}{2\omega(4\omega^2 - 1)} + (3 - 2\omega) \right| = \frac{|\epsilon^3|}{32(2\omega - 1)(\omega - 1)\omega} \qquad (32)$$

(see Fig. 3). For

$$\left| \frac{\epsilon^2}{2\omega(4\omega^2 - 1)} + (3 - 2\omega) \right| \neq \frac{|\epsilon^3|}{32\omega(2\omega - 1)(\omega - 1)}$$

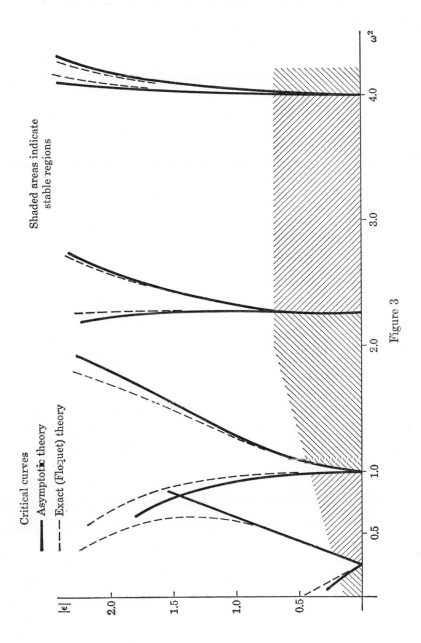

Figure 3

Critical curves
—— Asymptotic theory
— — Exact (Floquet) theory

Shaded areas indicate
stable regions

233

Eqs. (31) lead to either

$$\frac{A}{A_0} = \sqrt{\frac{a_2 + b_2 \cos \Phi_0}{a_2 + b_2 \cos \Phi}} \tag{33}$$

or

$$\frac{A}{A_0} = \sqrt{\frac{a_2 + b_2 \cos \Phi}{a_2 + b_2 \cos \Phi_0}} \tag{34}$$

with Φ given by (26) or (27), in which a_1 and b_1 have been replaced throughout by

$$a_2 = \frac{\epsilon^2}{2\omega(2\omega + 1)(2\omega - 1)} + (3 - 2\omega)$$

and

$$b_2 = \frac{\epsilon^3}{32\omega(2\omega + 1)(\omega - 1)}$$

respectively.   Along the critical curves (32), periodic solutions correspond to trivial solutions, and again there are nonperiodic bounded and unbounded solutions given by (33) and (34).   In this case we find that the fundamental frequency of the periodic solutions is given by $\omega - d\theta/dt = \omega - (2\omega - 3)/2 = \frac{3}{2}$, so that the least period is $4\pi/3$.   Of course, 4π is also a period, and hence these periodic solutions correspond to Mathieu functions.

The process may be continued to obtain higher-order approximations without essential modifications and, at each step, a new segment of the resonance pattern is revealed.   However, the labor involved soon becomes excessive.   The stability properties of the Mathieu equation (1), as inferred from the asymptotic solution (2), are indicated in Fig. 3.   The critical curves separate the resonant and nonresonant cases of (1).   Though the asymptotic character of (2) restricts any application to small $|\epsilon|$, the over-all qualitative characteristics of the stability regions in Fig. 3 are remarkably similar to those produced by the exact (Floquet) theory.   Well-known amplitude and frequency modulations in the principal part of the solution (for parameter pairs (ϵ,ω) near, but off the critical curves) are here reflected in the various closed form expressions for A and Φ.

EXERCISES

1. Explain the occurrence of periodic solutions of (1) for parameter pairs (ϵ,ω) "everywhere dense" in the shaded (stable) regions of the $\epsilon\omega$ plane in Fig. 3.   The periods 2π and 4π do not, however, occur.

2. Explain in detail the relationship between the resonant and nonresonant solutions obtained by the asymptotic method.   In particular, explain how they merge as the point (ϵ,ω) moves away from the critical curves.   What modifications need to be made in the "reduction" of the variational equations in order that the resonant and nonresonant solutions merge?

3. Apply the traditional method of "variation of parameters" to a study of (1).   Assume a solution of the form

$$u = A \cos (\omega t - \theta) \qquad du/dt = -\omega A \sin (\omega t - \theta)$$

in which each of A and θ is a variable.   Show that the equations of variation for A and θ assume the forms

$$\frac{d\theta}{dt} = \frac{\epsilon}{2\omega^2} \cos t + \frac{\epsilon}{4\omega^2} \cos [(2\omega - 1)t - 2\theta]$$

$$+ \frac{\epsilon}{4\omega^2} \cos [(2\omega + 1)t - 2\theta]$$

$$\frac{1}{A} \frac{dA}{dt} = -\frac{\epsilon}{2} \sin [(2\omega - 1)t - 2\theta] - \frac{\epsilon}{2} \sin [(2\omega + 1)t - 2\theta]$$

Note that this procedure is perfectly general and (1) is equivalent to the above system.   The motivation behind the "variation of parameters" method is the expectation that a solution of (1) for $\epsilon \neq 0$ (but for $|\epsilon|$ small) is very nearly a simple harmonic of the form $A \cos (\omega t - \theta)$, which it would be if $\epsilon = 0$.   The perturbations induced by the ϵ term in (1) are then expected to be reflected in "slow" changes of the amplitude A and phase θ of the near harmonic.   This expectation is borne out by the form of the above system since the right-hand members are each proportional to ϵ.   However, the integration of this system is not a simple matter. Using the method of "averaging," one replaces the right-hand members by their mean values with respect to t (on the premise that the rapid oscillations do not contribute to the slow changes which occur in θ and A). This reduces each of the right-hand members to zero (the zero function) provided $\omega \neq \frac{1}{2}$. This is appropriate so long as $|2\omega - 1|$ is not small. When it is, the corresponding terms no longer represent rapid oscillations. A little insight into the matter is obtained if one first introduces the auxiliary variable

$$\Phi = (1 - 2\omega)t + 2\theta$$

Show that the system then becomes

$$\frac{d\Phi}{dt} = 1 - 2\omega + \frac{\epsilon}{\omega^2} \cos t + \frac{\epsilon}{2\omega^2} \cos \Phi + \frac{\epsilon}{2\omega^2} \cos (2t - \Phi)$$
$$\frac{1}{A} \frac{dA}{dt} = \frac{\epsilon}{2} \sin \Phi + \frac{\epsilon}{2} \sin (2t - \Phi)$$

Unless ω is nearly $\frac{1}{2}$, Φ does not vary slowly. But if ω *is* nearly $\frac{1}{2}$, Φ not only varies slowly, but the method of averaging yields the nontrivial approximations

$$\frac{d\Phi}{dt} = 1 - 2\omega + \frac{\epsilon}{2\omega^2} \cos \Phi$$
$$\frac{1}{A} \frac{dA}{dt} = \frac{\epsilon}{2} \sin \Phi$$

This is the system (13), (14). Using the method of averaging, show how the other critical (resonance) cases arise in the variation of parameters version.

3. *The Free Oscillations of the van der Pol Equation*

Consider the van der Pol equation

$$\frac{d^2x}{dt^2} + x = \mu(1 - x^2) \frac{dx}{dt} \tag{35}$$

for $\mu > 0$ and an asymptotic solution of the form

$$x = A \cos (t - \theta) + \mu x_1 + \mu^2 x_2 + \cdots + \mu^N x_N \qquad (36)$$

where each of A, θ, x_1, . . . , x_N is a variable.   The principal part of the solution is chosen as a near harmonic, so as to reflect the basic characteristics of the solution for small μ.   For $\mu = 0$, of course, the amplitude A and phase θ are each constant.   When (36) is substituted into (35), one obtains the equation

$$\left[2A \frac{d\theta}{dt} + \frac{d^2 A}{dt^2} - A \left(\frac{d\theta}{dt} \right)^2 \right] \cos (t - \theta)$$
$$+ \left[A \frac{d^2 \theta}{dt^2} + 2 \frac{dA}{dt} \frac{d\theta}{dt} - 2 \frac{dA}{dt} \right] \sin (t - \theta)$$
$$+ \mu \left(\frac{d^2 x_1}{dt^2} + x_1 \right) + \mu^2 \left(\frac{d^2 x_2}{dt^2} + x_2 \right) + \cdots + \mu^N \left(\frac{d^2 x_N}{dt^2} + x_N \right)$$
$$= - \frac{\mu}{4} A (4 - A^2) \sin (t - \theta) + \frac{\mu}{4} A^3 \sin 3(t - \theta) + 0(\mu^2)$$
$$(37)$$

where $0(\mu^2)$ denotes terms of order μ^2, μ^3, ...   For $\mu = 0$, (37) implies, appropriately, that each of A and θ is a constant.   Considering only terms through first order in μ, we are led to the variational equations

$$2A \frac{d\theta}{dt} + \frac{d^2 A}{dt^2} - A \left(\frac{d\theta}{dt} \right)^2 = 0$$
$$A \frac{d^2 \theta}{dt^2} + 2 \frac{dA}{dt} \frac{d\theta}{dt} - 2 \frac{dA}{dt} = - \frac{\mu}{4} A (4 - A^2) \qquad (38)$$

and the perturbational equation

$$\frac{d^2 x_1}{dt^2} + x_1 = \frac{1}{4} A^3 \sin 3(t - \theta) \qquad (39)$$

Again it is sufficient to obtain solutions accurate to first order (in μ), and so we consider, instead of (38),

$$\frac{d\theta}{dt} = 0$$
$$\frac{dA}{dt} = \frac{\mu}{8} A (4 - A^2) \qquad (40)$$

Thus to first order in μ, θ remains constant and A is determined immediately as

$$A^2 = \frac{4}{1 + (4/A_0^2 - 1)e^{-\mu t}} \qquad (41)$$

where A_0^2 denotes an arbitrary (nontrivial) initial value for A^2. In every case, $A^2 \to 4$ (i.e. $|A| \to 2$) as $t \to \infty$. To first order, (39) is satisfied by $x_1 = -(A^3/32) \sin 3(t - \theta)$ and the first-order solution becomes

$$x = A \cos (t - \theta) - \mu \frac{A^3}{32} \sin 3(t - \theta) \qquad (42)$$

with θ constant and A given by (41). When (42) is used, the second-order terms on the right in (37) are then determined from the right-hand member of (35). It is also necessary to evaluate the term $\mu(d^2x_1/dt^2 + x_1)$ on the left in (37), to second order in μ. Considering terms through second order in μ, we are led to the variational equations

$$2A \frac{d\theta}{dt} + \frac{d^2A}{dt^2} - A \left(\frac{d\theta}{dt}\right)^2 = \frac{\mu^2 A^5}{128}$$

$$+ \frac{\mu^2 A}{32} (4 - 3A^2)(4 - A^2) \qquad (43)$$

$$A \frac{d^2\theta}{dt^2} + 2 \frac{dA}{dt} \frac{d\theta}{dt} - 2 \frac{dA}{dt} = - \frac{\mu}{4} A(4 - A^2)$$

and the perturbational equation

$$\frac{d^2x_2}{dt^2} + x_2 = \frac{5}{128} A^5 \cos 5(t - \theta) + \frac{A^3(A^2 + 8)}{128} \cos 3(t - \theta)$$

Using the first-order solution (40), we may reduce the variational equations (43) to the forms

$$\frac{d\theta}{dt} = \frac{\mu^2 A^4}{256} + \frac{\mu^2}{128} (4 - 3A^2)(4 - A^2)$$

$$\frac{dA}{dt} = \frac{\mu A}{8} (4 - A^2) \qquad (44)$$

which retain accuracy to second order in μ. The second of these is as in (40), and hence the amplitude A is given correctly to *second* order in μ by (41). The first equation in (44) may be expressed in the form

$$\frac{d\theta}{dt} = \frac{\mu^2}{16} + \frac{\mu^2}{256}(A^4 - 16) - \frac{\mu^2}{128}(4 - 3A^2)(4 - A^2)$$

$$= \frac{\mu^2}{16} + \frac{\mu^2}{256}(4 - 7A^2)(4 - A^2)$$

and thence, upon using the second, in the form

$$d\theta = \frac{\mu^2}{16}dt + \frac{\mu}{32A}(4 - 7A^2)\,dA$$

Thus

$$\theta = \theta_0 + \frac{\mu^2}{16}t \pm \frac{\mu}{8}\ln|A| - \frac{7\mu}{64}A^2 \qquad (45)$$

with θ_0 constant. The perturbational term x_2 is determined as

$$x_2 = -\frac{5}{3072}A^5\cos 5(t - \theta) - \frac{A^3}{1024}(A^2 + 8)\cos 3(t - \theta)$$

and the second-order solution becomes

$$x = A\cos(t - \theta) - \mu\frac{A^3}{32}\sin 3(t - \theta)$$

$$- \frac{5\mu^2}{3072}A^5\cos 5(t - \theta) - \frac{\mu^2 A^3}{1024}(A^2 + 8)\cos 3(t - \theta)$$

with the phase θ given by (45) and the amplitude A by (41). Higher-order solutions can, of course, be obtained by straightforward, though increasingly more tedious, calculations. One observes that no resonance phenomena have been (or will be) encountered.

EXERCISES

1. Plot the principal part of a solution (36) for various initial amplitudes A_0.

2. If the initial amplitude A_0 is 2, (36) becomes an expansion of the periodic solution of (35). Show that in this special case,

the present method is equivalent to the Poincaré expansion procedure introduced in Chap. 3. In particular, show that (45) yields the second-order correction to the fundamental frequency of the periodic solution. Explain how (45) defines an "asymptotic phase" for each nonperiodic (transient) solution.

3. Obtain the third-order solution (36). Note, in particular, that the first equation in (44) is valid to order μ^3, but that the second now requires modification.

4. Use the variation of parameters method and subsequently the method of averaging in a study of the solutions of (35).

4. *The Forced Oscillations of the van der Pol Equation*

We first consider the equation

$$\frac{d^2x}{dt^2} + x = \mu(1 - x^2)\frac{dx}{dt} + \mu k \cos \lambda t \qquad (46)$$

where each of k and $\lambda > 0$ is a fixed constant and the constant $\mu > 0$ is small. Because of the factor μ, the forcing function in (46) is "soft," and so we here anticipate an asymptotic solution in the form

$$x = A \cos (t - \theta) + \mu x_1 + \mu^2 x_2 + \cdots + \mu^N x_N \qquad (47)$$

with a principal part reflecting the basic characteristics of the free oscillations. The "hard" forced case will be considered subsequently. When (47) is substituted into (46), one obtains the equation

$$\left[2A \frac{d\theta}{dt} + \frac{d^2A}{dt^2} - A \left(\frac{d\theta}{d}\right)^2 \right] \cos (t - \theta)$$

$$+ \left[A \frac{d^2\theta}{dt^2} + 2 \frac{dA}{dt} \frac{d\theta}{dt} - 2 \frac{dA}{dt} \right] \sin (t - \theta)$$

$$+ \mu \left(\frac{d^2x_1}{dt^2} + x_1\right) + \cdots + \mu^N \left(\frac{d^2x_N}{dt^2} + x_N\right) = \mu k \cos \lambda t$$

$$- \frac{\mu}{4} A(4 - A^2) \sin (t - \theta) + \frac{\mu}{4} A^3 \sin 3(t - \theta) + 0(\mu^2) \qquad (48)$$

If the input frequency λ is appreciably different from unity, then we obtain for the first-order variational equations

$$\frac{d\theta}{dt} = 0$$

$$\frac{dA}{dt} = \frac{\mu}{8} A (4 - A^2)$$

and for the first-order perturbational equation

$$\frac{d^2 x_1}{dt^2} + x_1 = \frac{A^3}{4} \sin 3(t - \theta) + k \cos \lambda t \qquad (49)$$

From these we obtain the first-order solution

$$x = A \cos (t - \theta) - \mu \frac{A^3}{32} \sin 3(t - \theta) + \frac{\mu k}{1 - \lambda^2} \cos \lambda t \quad (50)$$

with θ constant and A given by (41). This solution contains both the forced and natural frequencies superimposed as though the system were linear. Since $|A|$ approaches 2, the natural response can be expected to dominate. This is not an unexpected result, of course, in view of the fact that (46) is "nearly" linear, and the forcing function is soft. However, as the forcing frequency λ approaches unity, the character of the solution is modified radically and a somewhat unexpected phenomenon takes place. As λ approaches unity, the forced response becomes more significant [this is already clear in (50)] but instead of a persistence of both the natural and the forced responses, the former becomes *entrained* by the latter. The result is a synchronization of output at the input frequency. Put another way, the soft forcing function in (46) commands a response at the input frequency only under a resonant condition with an input frequency nearly equal to the natural frequency. Expressed in this way, the entrainment phenomenon is perhaps not quite so unexpected.

For λ nearly unity, we remove the nearly resonant term in (49), express it in the form

$$\mu k \cos \lambda t = \mu k \cos [t(\lambda - 1) + \theta)] \cos (t - \theta)$$
$$- \mu k \sin [t(\lambda - 1) + \theta] \sin (t - \theta)$$

and combine the result with the other fundamental harmonics. The variational equations are then modified according to

$$A \frac{d\theta}{dt} = \frac{\mu k}{2} \cos [t(\lambda - 1) + \theta]$$
$$\frac{dA}{dt} = \frac{\mu}{8} A(4 - A^2) + \frac{\mu k}{2} \sin [t(\lambda - 1) + \theta] \tag{51}$$

and the perturbational equation becomes merely (49) with the nearly resonant term removed.   The study of the variational system (51) is facilitated by introducing the rectangular coordinates

$$a = A \cos [t(\lambda - 1) + \theta]$$
$$b = A \sin [t(\lambda - 1) + \theta] \tag{52}$$

in lieu of the polar-type coordinates (A,θ).   The ab plane is referred to as the *van der Pol plane*.   In terms of the rectangular van der Pol coordinates (a,b), the system (51) becomes

$$\frac{da}{dt} = \frac{\mu}{8} (4 - A^2)a + (1 - \lambda)b$$
$$\frac{db}{dt} = \frac{\mu}{8} (4 - A^2)b - (1 - \lambda)a + \frac{\mu}{2} k \tag{53}$$

where $A^2 = a^2 + b^2$.   A critical point (a_0,b_0) of the system (53) satisfies the algebraic equations

$$\frac{\mu}{8} (4 - A_0{}^2)a_0 + (1 - \lambda)b_0 = 0$$
$$(1 - \lambda)a_0 - \frac{\mu}{8} (4 - A_0{}^2)b_0 = \frac{\mu}{2} k \tag{54}$$

which may also be expressed in the forms

$$a_0 = -\frac{(k/2)(1 - \lambda)/\mu}{\frac{1}{64}(4 - A_0^2)^2 + [(1 - \lambda)/\mu]^2}$$

$$b_0 = \frac{(k/16)(4 - A_0^2)}{\frac{1}{64}(4 - A_0^2)^2 + [(1 - \lambda)/\mu]^2}$$

(55)

Upon squaring and adding in (54), we obtain also the useful relation

$$A_0^2 \left[\frac{1}{64}(4 - A_0^2)^2 + \left(\frac{1 - \lambda}{\mu}\right)^2 \right] = \frac{1}{4}k^2$$

(56)

For each nonnegative root A_0^2 of (56), Eqs. (55) define a critical point (a_0, b_0) of (53). The nature of a critical point is generally reflected in the linear approximation near that critical point. This is the linear system with the coefficient matrix [Jacobian matrix with respect to (a,b) of the right-hand members of (53) evaluated at (a_0, b_0)]

$$\begin{pmatrix} \frac{\mu}{8}(4 - A_0^2 - 2a_0^2) & 1 - \lambda - \frac{\mu}{4} a_0 b_0 \\ -1 + \lambda - \frac{\mu}{4} a_0 b_0 & \frac{\mu}{8}(4 - A_0^2 - 2b_0^2) \end{pmatrix}$$

The determinant is

$$\Delta = \mu^2 \left[\frac{1}{64}(4 - A_0^2)(4 - 3A_0^2) + \left(\frac{1 - \lambda}{\mu}\right)^2 \right]$$

and the trace is

$$\Omega = \frac{\mu}{2}(2 - A_0^2)$$

In the variables A_0^2 and $(1 - \lambda)/\mu$, the locus $\Delta = 0$ is an ellipse, the interior of which corresponds to $\Delta < 0$. The corresponding critical points of (53) are saddle points. Exterior to the ellipse, $\Delta > 0$, and the corresponding critical points are stable if $\Omega > 0$ and unstable if $\Omega < 0$. The locus $\Omega = 0$ is the line $A_0^2 = 2$.

Figure 4 depicts the complete pattern distinguishing, in addition, those regions which correspond to foci and those which correspond to nodes.

Now a critical point (a_0, b_0) of the variational system (53) corresponds to a periodic solution of (46). The amplitude of the principal part (to first order in μ) is $|A_0|$ and the frequency is

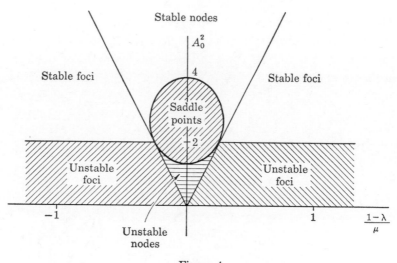

Figure 4

[according to (52)] given by $1 - d\theta/dt = 1 - (1 - \lambda) = \lambda$, i.e., the impressed frequency. The complete solution (to first order in μ) becomes

$$x = A_0 \cos \lambda t - \mu \frac{A_0{}^3}{32} \sin 3\lambda t \tag{57}$$

The orbital stability characteristics of one of these periodic solutions are the same as the Liapunov stability characteristics of the corresponding critical point of (53).

It is of some interest perhaps to consider the nontrivial solutions of (53). We are not so fortunate in this case as to be able to exhibit them in closed form. Hence, we shall be content here with a few observations of a rather general nature and then a look at some special cases (see Figs. 5, 6, and 7 for illustrations of a few of the possibilities). From (53) we have immediately

$$\frac{1}{2}\frac{dA^2}{dt} = a\frac{da}{dt} + b\frac{db}{dt} = \frac{\mu}{8}(4 - A^2)A^2 + \frac{\mu}{2}kb$$

$$= -A^2\left[\frac{\mu}{8}(A^2 - 4) - \frac{\mu k}{2}\frac{b}{A^2}\right]$$

Thus, for A^2 large, dA^2/dt is negative, and this means that the distant solutions in the ab plane are necessarily directed inward (see Fig. 5). If we introduce in the ab plane the polar angle Φ in the usual manner, i.e., $a = A\cos\Phi$, $b = A\sin\Phi$, then accord-

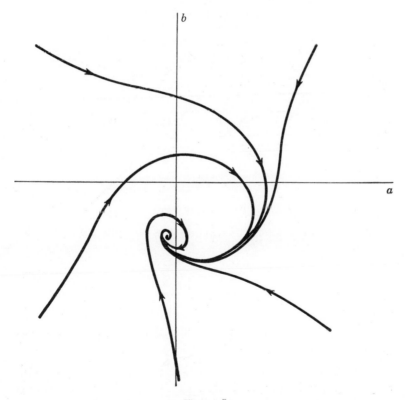

Figure 5

ing to (52), we have $\Phi = t(\lambda - 1) + \theta$, and so the asymptotic solution (47) assumes the form

$$x = A \cos (\lambda t - \Phi) + \mu x_1 + \cdots + \mu^N x_N \qquad (58)$$

Thus, the deliberate motion in the van der Pol plane given by (53) is manifest directly in (58) by slow variations in the ampli-

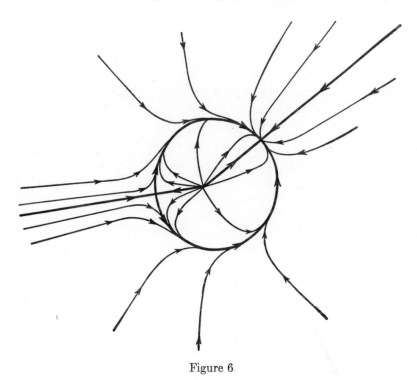

Figure 6

tude A and phase Φ, the latter now defined relative to the impressed frequency. The above observation implies that all nonperiodic solutions exhibit deliberate, possibly somewhat sporadic but uniformly bounded, amplitude modulations (see Fig. 6). Frequency modulations occur if the corresponding van der Pol trajectory persists in encircling the origin of the van der Pol plane (see Fig. 7). In such a case, we say that Φ has a *secular part* (a time-like part). When the quantity $|1 - \lambda|/\mu$ is relatively large, (56) possesses exactly one positive root $A_0{}^2$. Unless k is also correspondingly large, this one positive root will be small and will lead to a critical point lying in the unstable region of

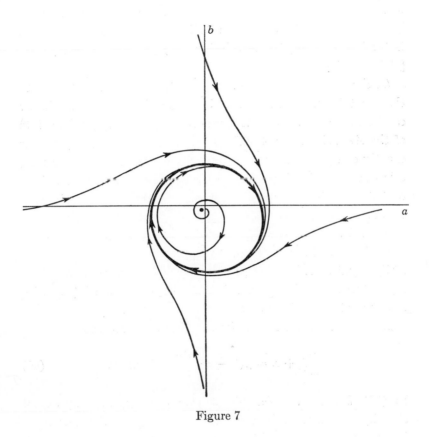

Figure 7

Fig. 4. By the Poincaré-Bendixson theorem, then, there will exist a periodic orbit of (53). It turns out that each nontrivial solution of (53) tends to the periodic one (see Fig. 7). In this fashion, the form (58) of the asymptotic solution merges with the form (50) for λ appreciably different from 1. A discussion of the exact correspondence between these two forms is an interesting

exercise, which is left to the student. The unique unstable
critical point encountered here leads to an unstable periodic
solution of (46) at the impressed frequency λ. This theoretically
interesting (practically, uninteresting) periodic solution was not
discussed in the nonresonant case of (50) because we more or less
ignored the trivial variational solution $A = 0$, $\theta = $ const. This
latter leads to a periodic solution (50) with the leading term
$[\mu k/(1 - \lambda^2)] \cos \lambda t$.

At the other extreme, we have $\lambda = 1$, and this corresponds to
the exact *linear* resonance case in (46). Here, in the nonlinear
case, we are concerned with those cases represented by the points
of the $A_0{}^2$ axis in Fig. 4. For k^2 exceeding $16/27$, (56) possesses
exactly one positive root $A_0{}^2$, and as k^2 increases without bound,
this one root also increases without bound. Thus the unique
root $A_0{}^2$ will for all large k^2 fall within the stable region of Fig. 4.
The corresponding periodic solution (58) is asymptotically orbitally stable. In fact, every solution necessarily approaches the
periodic one. Similar results hold for all λ near but not necessarily equal to 1. That is, for all large k^2, there exists a unique
asymptotically orbitally stable periodic solution of (46) with
frequency λ, to which every solution converges.

We rephrase these results for the "hard" forced case

$$\frac{d^2x}{dt^2} + x = \mu(1 - x^2)\frac{dx}{dt} + F \cos \lambda t \qquad (59)$$

In (59), k becomes F/μ. For λ nearly unity, we may expect
a unique periodic orbitally stable solution at the impressed
frequency λ and to which every solution converges. For λ
appreciably different from unity, we may expect a unique asymptotically orbitally stable solution which exhibits both the natural
and the impressed frequencies and to which every solution converges, save a single unstable periodic solution at the impressed
frequency λ. One encounters no difficulty in exhibiting these
results directly. We assume an asymptotic solution with the

principal part

$$A \cos (t - \theta) + \frac{F}{1 - \lambda^2} \cos \lambda t \qquad (60)$$

which reflects the basic characteristics of the linear "hard" forced case. Then for λ appreciably different from 1, we obtain (to first order in μ) the variational equations

$$\frac{d\theta}{dt} = 0$$
$$\frac{dA}{dt} = \frac{\mu A}{8}\left[1 - \frac{A^2}{4} - \frac{F^2}{2(1 - \lambda^2)^2}\right] \qquad (61)$$

The first-order perturbational equation contains quite a number of terms, one of which becomes resonant for $\lambda = 1$. We shall not be concerned with resonance here, however. It is the second equation in (61) which is of interest. Of course, it is but a simple matter to integrate this equation. We leave this matter as an exercise for the student. Excepting the null solution $A^2 = 0$, it is clear that every solution A^2 either tends to zero, in the event $F^2/2(1 - \lambda^2)^2 \geq 1$, or tends to $4[1 - F^2/2(1 - \lambda^2)^2]$, in the event $F^2/2(1 - \lambda^2)^2 < 1$. In the former circumstance, the "natural" response portion of the solution fades out completely, so that only the impressed frequency λ remains in the steady-state oscillation. In the latter circumstance, both the natural and impressed frequencies will persist into the steady-state oscillation. Thus, in the hard forced case, the entrainment of the natural frequency by the impressed frequency will occur throughout a rather wide range of input frequencies λ. It should be noted that the rapidity with which A^2 tends to zero for $F^2/2(1 - \lambda^2)^2 \geq 1$ increases markedly with increases in the quantity $F^2/2(1 - \lambda^2)^2$. Thus, the rapidity with which the natural response portion fades out also increases markedly with either increases in the hardness of the forcing function or else decreases in the separation of the natural and impressed frequencies. The absolute value of the difference between the natural and impressed frequencies is referred to as the *detuning*.

EXERCISES

1. Assume a solution of (46) in the form

$$x = a \cos \lambda t + b \sin \lambda t \qquad \frac{dx}{dt} = -\lambda a \sin \lambda t + \lambda b \cos \lambda t$$

where each of a and b is a variable. Derive the variation of parameters equivalent of (46) in terms of a and b. Use the method of averaging (over the period $2\pi/\lambda$) to reduce the variational system to autonomous form. Note that the result is the system (53). This is referred to as the *van der Pol method*. The analysis of the variational system is referred to as the *method of Andronow and Witt*.

2. Derive the variation of parameters equivalent of (46) for a solution of the form

$$x = A \cos (\lambda t - \Phi) \qquad \frac{dx}{dt} = -\lambda A \sin (\lambda t - \Phi)$$

where each of A and Φ is a variable, and then use the method of averaging to reduce the variational system to an autonomous form. Discuss the relationship between this and the approach in Exercise 1. This is referred to as the *method of the first approximation of Krylov and Bogoliubov*.

3. Discuss in detail the correspondence between (50) and (58). What modification in the variational equations (51) is required in order that the resonant and nonresonant cases merge? Show that any solution of the system

$$A \frac{d\theta}{dt} = \frac{\mu k}{1 + \lambda} \cos [t(\lambda - 1) + \theta]$$

$$\frac{dA}{dt} = \frac{\mu}{8} A(4 - A^2) + \frac{\mu k}{1 + \lambda} \sin [t(\lambda - 1) + \theta]$$

$$\frac{d^2 x_1}{dt^2} + x_1 = \frac{A^3}{4} \sin 3(t - \theta)$$

satisfies Eq. (48) to first order in μ for *all* values of λ. Show that for λ nearly unity, this system leads to (58), while for λ appreciably different from unity, it leads to (50). Use this system to discuss in detail the correspondence between the three forms (50), (58), and (60).

4. Discuss the asymptotic expansions, to first order in μ, of the solution of (59) for $\lambda \neq 1$. In particular, verify (61).

5. *The Forced Oscillations of the Duffing Equation*

For the Duffing equation

$$\frac{d^2x}{dt^2} + x = \beta x^3 + \beta F_0 \cos \lambda t \tag{62}$$

with $|\beta|$ small, we consider an asymptotic solution of the form

$$x = A \cos (t - \theta) + \beta x_1 + \beta^2 x_2 + \cdots + \beta^N x_N \tag{63}$$

With λ appreciably different from unity, we obtain, to first order in β, the variational equations

$$\frac{d\theta}{dt} = \frac{3}{8} \beta A^2$$

$$\frac{dA}{dt} = 0$$

and the perturbational equation

$$\frac{d^2x_1}{dt^2} + x_1 = \frac{A^3}{4} \cos 3(t - \theta) + F_0 \cos \lambda t \tag{64}$$

The solution, to first order in β, becomes

$$x = A \cos (t - \theta) - \frac{\beta A^3}{32} \cos 3(t - \theta) + \frac{\beta F_0}{1 - \lambda^2} \cos \lambda t \tag{65}$$

with A constant and $\theta = \theta_0 + 3\beta A^2 t/8$. Thus, in general, both the natural and impressed frequencies appear in the solution.

The resonance case, λ near 1, is of more interest. The resonant term in (64) is removed, expressed as the sum of two nearly harmonic $\cos (t - \theta)$ and $\sin (t - \theta)$ terms, and then incorporated in the variational equations.   The following system results

$$\frac{d\theta}{dt} = \frac{3}{8}\beta A^2 + \frac{\beta F_0}{2A} \cos\left[(\lambda - 1)t + \theta\right]$$
$$\frac{dA}{dt} = \frac{\beta F_0}{2} \sin\left[(\lambda - 1)t + \theta\right]$$

These equations become

$$\frac{d\Phi}{dt} = \frac{3}{8}\beta A^2 + \lambda - 1 + \frac{\beta F_0}{2A} \cos\Phi$$
$$\frac{dA}{dt} = \frac{\beta F_0}{2} \sin\Phi \tag{66}$$

with $\Phi = (\lambda - 1)t + \theta$.   The singular solutions of (66) correspond to $\sin\Phi = 0$ and to the equation

$$\frac{3}{8}A_0^2 + \frac{\lambda - 1}{\beta} \pm \frac{F_0}{2A_0} = 0 \tag{67}$$

for the amplitude A_0.   Corresponding to each real root A_0 of
(67), there is a periodic solution of (62) of frequency

$$1 - \frac{d\theta}{dt} = 1 - (1 - \lambda) = \lambda$$

i.e., the impressed frequency, with a principal part of amplitude
$|A_0|$, to first order in β.   Equation (67) is merely a version of the
equation for the bifurcation amplitudes of the Duffing equation
introduced in Chap. 7 (see Eq. (18) and Fig. 2 of Chap. 7).

In this case, we are able to obtain the nontrivial solutions of
(66) in closed form.[1]   From this closed form solution we can, of
course, determine, among other things, the stability of each
of the singular solutions.   We then will have determined the
stability of the corresponding periodic solutions of frequency λ
of (62).

[1] The author is indebted to Mr. Steve M. Yionoulis for suggesting the
general integrability of the system (66).

Upon dividing the first equation in (66) by the second, there results the differential form

$$\frac{F_0 A}{2} \sin \Phi \, d\Phi - \left[\frac{3}{8} A^3 + \left(\frac{\lambda - 1}{\beta} \right) A + \frac{F_0}{2} \cos \Phi \right] dA = 0$$

which is exact. The general integral

$$F_0 A \cos \Phi + \left(\frac{\lambda - 1}{\beta} \right) A^2 + \frac{3}{16} A^4 = c$$

where c is a constant, is obtained immediately. The singular solutions appear for

$$c = \pm F_0 A_0 + \left(\frac{\lambda - 1}{\beta} \right) A_0^2 + \frac{3}{16} A_0^4 = \pm \frac{F_0 A_0}{2} - \frac{3}{16} A_0^4$$

$$\tag{68}$$

where A_0 is a real root of (67), and the $(\pm)$ sign is as in (67). It is convenient to map the solution curves into the van der Pol plane with $a = A \cos \Phi$ and $b = A \sin \Phi$. The system (66) may be expressed in the form

$$\frac{1}{\beta} \frac{da}{dt} = - \left(\frac{3}{8} A^2 + \frac{\lambda - 1}{\beta} \right) b$$
$$\frac{1}{\beta} \frac{db}{dt} = \left(\frac{3}{8} A^2 + \frac{\lambda - 1}{\beta} \right) a + \frac{F_0}{2}$$

$$\tag{69}$$

with the singular solutions occurring for $b = 0$ and $a = \pm A_0$, where A_0 is a real root of (67). The solution curves in the ab plane become

$$A^2 \left[\frac{3}{16} A^2 + \left(\frac{\lambda - 1}{\beta} \right) \right] = c - F_0 a \tag{70}$$

where, of course, $A^2 = a^2 + b^2$ is the square of the length of the radius vector. In particular, (70) is a quartic in a and b. It is clear that all solutions are bounded, since the term $(3/16) A^4$ completely dominates all others for large A.

For $(\lambda - 1)/\beta$ positive, (67) possesses exactly one real root, and the quartic curves (70) form a family of concentric closed paths centered about the corresponding singular point $a = \pm A_0$, $b = 0$. These are best described by considering the locus

$$\left(\frac{3}{8} A^2 + \frac{\lambda - 1}{\beta}\right) a = -\frac{F_0}{2} \tag{71}$$

which, according to the second equation in (69), is the isocline of horizontal tangents. The isocline (71) is illustrated in Fig. 8. Its intercept with the a axis is the singular point $a = \pm A_0$, $b = 0$. In fact, the a axis is the isocline of vertical tangents. For c large in (70), the trajectory is very nearly the circle

$$(\tfrac{3}{16}) A^4 = c$$

As c decreases, the trajectory becomes distorted somewhat but remains a closed path. For $c = 0$, the solution passes through the origin and as c continues to decrease, the trajectory approximates an ellipse with center at the singular point. The singular point is reached when c has decreased to the value in (68), which in this case will be negative. For c less than the value in (68), (70) will not yield a real trajectory.

For $(\lambda - 1)/\beta$ negative, the picture may be considerably modified. If F_0 is sufficiently large, the locus (71) will be qualitatively similar to that in Fig. 8, and there will be but one singular point as before. But if F_0 is a small positive quantity, then the locus (71) will possess two branches. It is clear that exterior to the circle $A^2 = 8(1 - \lambda)/3\beta$, there is always a branch of the locus which is qualitatively similar to that in Fig. 8. However, for F_0 small, inside this circle there will be a second branch which forms an oval and lies on the opposite side of the b axis. The limitation as to the size of F_0 is clear, since the left-hand member of (71) vanishes everywhere on the circle $A^2 = -8(\lambda - 1)/3\beta$, as well as along the b axis, and is thus bounded within. The com-

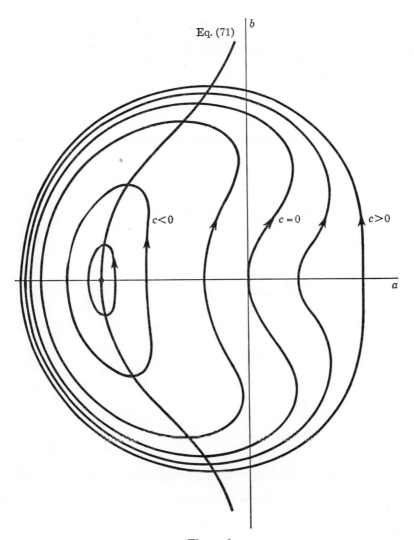

Figure 8

plete locus (71) for a small positive F_0 is illustrated in Fig. 9.   The circle $A^2 = 8(1 - \lambda)/3\beta$ is a branch of the isocline of vertical tangents.   The remaining branch is, of course, the a axis.   In this case, (67) possesses three real roots which are represented by the intercepts of the locus (71) with the a axis.   There are two centers and a saddle point arranged in an unusual pattern.   The com-

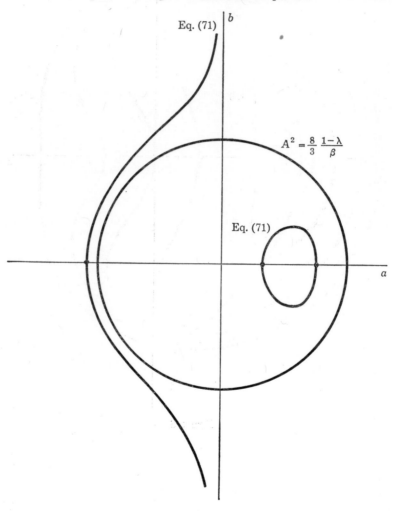

Figure 9

plete trajectory pattern is illustrated in Fig. 10.   The saddle
point and the right-hand center coalesce and then vanish along
with the oval branch of (71), as F_0 reaches and then exceeds the
appropriate extremum in (71).   The distant trajectories are very

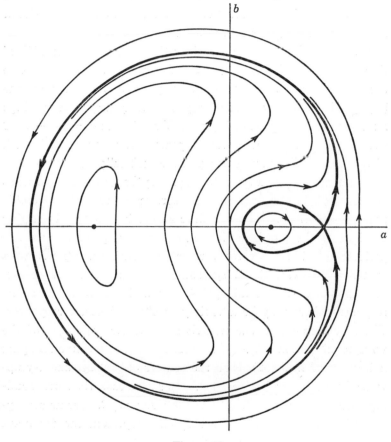

Figure 10

nearly the circles $(\frac{3}{16})A^4 = c$, as before.   With small F_0 and
as c decreases, a value is reached which corresponds to the right-
hand center.   This is the largest of the three possible values in
(68).   For somewhat smaller values of c, there are two periodic

solutions, one large, one small.   As c continues to decrease and becomes negative, a second of the values in (68) is reached.   This corresponds to the saddle point and two looping separatrices. For still smaller c, a single (new and somewhat distorted) periodic solution exists, and as c decreases further, the final (minimum) value in (68) is reached.   This corresponds to the left-hand center.   Real solutions do not exist for smaller c.

It is of some interest to consider the implications for (62) of these results.   With an impressed frequency λ nearly equal to the natural frequency, there exist periodic responses at the impressed frequency.   The amplitude-frequency relationships for these are illustrated in Fig. 2 of Chap. 7 with $\sigma = 2(1 - \lambda)/\beta$ and $|a_0| = |A_0|$.   Those points appearing along an upper branch of a constant F_0 curve correspond to orbitally stable periodic solutions.   Those points appearing along the lower-most portions of the parabola like branches also correspond to orbitally stable periodic solutions.   Those points appearing along the upper portions of the parabola like branches correspond to unstable periodic solutions.   Of the former two, each corresponds to a center of the variational system (69).   The latter, of course, corresponds to the saddle point.   In addition to the periodic responses, there may be a variety of almost periodic responses. These correspond to the nontrivial (periodic) solutions of the variational system (69).   If a small amount of positive damping is introduced into (62), then the centers of (69) become asymptotically stable foci (see Fig. 11).   In such a case, the corresponding periodic responses of (62) at the frequency λ become asymptotically orbitally stable, and they each represent the dominant part of a steady-state response.

Finally, let us illustrate briefly still another source of interesting nonlinear resonance phenomena.   For the equation

$$\frac{d^2x}{dt^2} + x = \beta x^3 + F_1 \cos \lambda_1 t + F_2 \cos \lambda_2 t \tag{72}$$

with two input frequencies λ_1 and λ_2 we assume an asymptotic solution of the form

$$x = A \cos (t - \theta) + \frac{F_1}{1 - \lambda_1{}^2} \cos \lambda_1 t + \frac{F_2}{1 - \lambda_2{}^2} \cos \lambda_2 t + \beta x_1$$
$$+ \beta^2 x_2 + \cdots + \beta^N x_N \quad (73)$$

Here we assume that $|\lambda_1|$ and $|\lambda_2|$ are each appreciably different from 1 and introduce into the expansion terms which represent the linearized responses to the "hard" forcing function in (72).

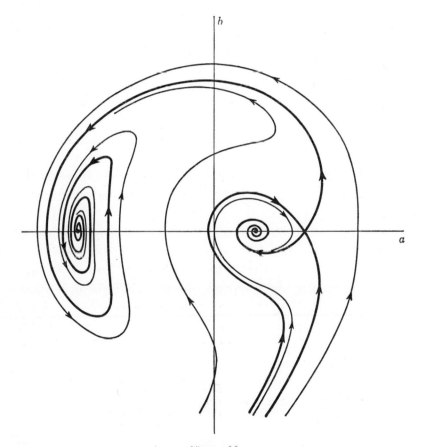

Figure 11

When (73) is substituted into (72), there appear a variety of possible resonant or near resonant terms. The frequencies $3\lambda_1$, $3\lambda_2$, $2\lambda_1 \pm \lambda_2$, $\lambda_1 \pm 2\lambda_2$, $2 \pm \lambda_1 - 2\,d\theta/dt$, $2 \pm \lambda_2 - 2\,d\theta/dt$, $1 \pm 2\lambda_1 - d\theta/dt$, $1 \pm 2\lambda_2 - d\theta/dt$ are present in first-order terms (in β). Should one (or more) of these in absolute value be nearly unity, the corresponding term (or terms) would produce resonance or near resonance in the first-order perturbational equation for x_1. It (or they) should be shifted to the variational equations. For example, if $2\lambda_1 - \lambda_2$ is nearly unity in absolute value, and no others, then we are led to the first-order variational equations

$$\frac{d\Phi}{dt} = \frac{3}{8}\beta A^2 + \frac{3}{4}\frac{\beta F_1^2}{(1-\lambda_1^2)^2} + \frac{3}{4}\frac{\beta F_2^2}{(1-\lambda_2)^2} + 2\lambda_1 - \lambda_2$$
$$+ \frac{\beta F_1^2 F_2}{8A(1-\lambda_1^2)^2(1-\lambda_2^2)}\cos\Phi \quad (74)$$
$$\frac{dA}{dt} = \frac{3}{8}\frac{\beta F_1^2 F_2}{(1-\lambda_1)^2(1-\lambda_2^2)}\sin\Phi$$

where $\Phi = (2\lambda_1 - \lambda_2 - 1)t + \theta$. This system possesses the same structure as (66) and so it is unnecessary to discuss it further.

In most cases, it is necessary to anticipate a variety of perturbational terms which have frequencies that are composed of linear combinations of the input frequencies λ_1 and λ_2 with integral coefficients. Some of these combination frequencies may lead to resonance or near resonance in the higher-order approximations. This is referred to as the *problem of small divisors* since if the resonant frequencies are not removed from the perturbational equation, they will lead to small divisors. In the present procedure, however, one merely shifts such potentially troublesome terms to the variational equations, and the small divisors never materialize.

EXERCISES

1. In (65), let $\lambda t = 3(t - \theta)$ and express the solution in terms of the impressed frequency λ. Note that the principal term of the response is a *subharmonic* of the input harmonic. Using (65),

obtain second-order resonant and nonresonant solutions of the Duffing equation (62). Note that resonance occurs for λ near 3. This is called *subharmonic resonance*.

2. Discuss the limit as $\beta \to 0$ of the periodic solutions of (62). Discuss the relationship between this and the bifurcation amplitudes for the Duffing equation introduced in Chap. 7. Note the sharp distinction between the above limiting procedure and its counterpart in Chap. 7.

3. Show how to use the integral (70) to express the solution of (66) in closed form as an elliptic integral.

4. Derive the system (74) and discuss its solutions. In particular, show that a singular solution of (74) corresponds to a periodic leading term in (73) of frequency $2\lambda_1 - \lambda_2$. This is called a *combination frequency* (or tone) because it is, in general, neither a subharmonic nor a superharmonic of either of the input frequencies.

5. Discuss the several first-order resonance possibilities in (72). Derive the variational equations, and discuss their solutions.

6. Use the variation of parameters method in conjunction with the method of averaging to derive (66).

7. Consider the soft-forced equation

$$\frac{d^2x}{dt^2} + x = \beta x^3 + \beta F_0 \cos \lambda_1 t + \beta F_0^* \cos \lambda_2 t$$

where each of the input frequencies λ_1 and λ_2 is nearly unity. Introduce, into the second order, nonautonomous variational system, two (suitable) auxiliary variables and express the variational system as a *third*-order autonomous system.

8. Discuss the forced oscillations of the slightly damped Duffing equation

$$\frac{d^2x}{dt^2} + x = \beta x^3 - \beta k \frac{dx}{dt} + \beta F_0 \cos \lambda t$$

In particular, verify the qualitative features of the solutions of the first-order variational system as illustrated in Fig. 11.

GENERAL REFERENCES

1. Bellman, Richard: "Stability Theory of Differential Equations," McGraw-Hill Book Company, Inc., New York, 1953.

2. Coddington, E. A., and N. Levinson: "Theory of Ordinary Differential Equations," McGraw-Hill Book Company, Inc., New York, 1955.

3. Lefschetz, Solomon: "Differential Equations: Geometric Theory," Interscience Publishers, Inc., New York, 1957.

4. Stoker, J. J.: "Nonlinear Vibrations," Interscience Publishers, Inc., New York, 1950.

5. Cesari, L.: "Asymptotic Behavior and Stability Problems in Ordinary Differential Equations," Springer-Verlag, Berlin, 1959.

6. Kaplan, Wilfred: "Ordinary Differential Equations," Addison-Wesley Publishing Company, Reading, Mass., 1958.

7. Minorsky, N.: "Introduction to Non-linear Mechanics," Edwards Bros., Inc., Ann Arbor, Mich., 1947.

8. Bellman, Richard: "An Introduction to Matrix Analysis," McGraw-Hill Book Company, Inc., New York, 1960.

9. McLachlan, N. W.: "Theory and Application of Mathieu Functions," Oxford University Press, New York, 1947.

10. Byrd, P. F., and M. D. Friedman: "Handbook of Elliptic Integrals for Physicists and Engineers," Springer-Verlag, Berlin, 1954.

INDEX